# TREE BY TREE

# TREE BY TREE

*Saving North America's Eastern Forests*

SCOTT J. MEINERS

COMSTOCK PUBLISHING ASSOCIATES
AN IMPRINT OF
CORNELL UNIVERSITY PRESS
ITHACA AND LONDON

First published 2023 by Cornell University Press

Library of Congress Cataloging-in-Publication Data

Names: Meiners, Scott J., 1970– author.
Title: Tree by tree : saving North America's eastern forests / Scott J. Meiners.
Description: Ithaca [New York] : Comstock Publishing Associates, an imprint of Cornell University Press, 2023. | Includes bibliographical references and index.
Identifiers: LCCN 2023003123 (print) | LCCN 2023003124 (ebook) | ISBN 9781501771262 (paperback) | ISBN 9781501771279 (pdf) | ISBN 9781501771286 (epub)
Subjects: LCSH: Trees—Ecology—Northeastern States. | Trees—Conservation—Northeastern States. | Forest declines—Northeastern States.
Classification: LCC QK477 .M38 2023 (print) | LCC QK477 (ebook) | DDC 582.160978—dc23/eng/20230209
LC record available at https://lccn.loc.gov/2023003123
LC ebook record available at https://lccn.loc.gov/2023003124

All botanical drawings are from C. S. Sargent, *The Silva of North America* (1892).

To all my teachers and students, with gratitude

No one can really know the forest without feeling the gentle influence of one of the kindliest and strongest parts of nature. From every point of view it is one of the most helpful friends of man. Perhaps no other natural agent has done so much for the human race and has been so recklessly used and so little understood.

Gifford Pinchot, first head of the United States Forest Service, *A Primer of Forestry*, 1899

# CONTENTS

# PREFACE

Tree by tree, we are losing species from our Eastern forests, and few have seemed to notice or to care. If unaltered, this loss is a genuinely frightening trajectory to find ourselves upon, with long-term implications for nature and our local and national economies. My personal life and professional career have been spent mainly in and around forests, so I have noticed these sad changes as they have occurred. There is an old saying that one cannot see the forest for the trees. This book will see the forest *and* the trees, as I strongly feel that understanding the forest can happen only when a person understands the trees that compose them.

Now, I must admit I unreservedly and unequivocally love trees. Not in the purely romantic way that one fantasizes about the "wilderness" experience or the "balance of nature." Instead, I love trees for what they are, in every way possible. They are beautiful in our landscapes, whether as a solitary tree in vast savannah grasslands, as open pine woodlands with a dense herbaceous understory, or as dense, shady forests with a carpet of spring wildflowers in bloom. I love walking up to a tree that I have never seen before and figuring out what it is, often to the annoyance of those around me. I love what you can do with wood. One of my favorite smells is that of a construction site, where the scent of the wood framing is strong with resinous pine. It is a shame that our homes do not continue to smell like that once they are finished. I love building things from wood, particularly furniture. There are few joys greater than conceiving a project, building it, and then watching as the finishing process magically brings out the wood's grain and character. I love chopping firewood for the internal warmth and sense of accomplishment it brings; I love the peaceful beauty of fires more. I love orchards for the patterns they make on the landscape and the fruit and nuts they produce. I love the diversity of leaf shapes, both naturally occurring and those bred by the horticultural industry. I love the variety of tree forms, from the shortest trees growing at a mountain's tree line, pruned like a bonsai by the wind and snow, to the tallest spreading versions that appear regally upon a sunset skyline.

Where did this love of trees originate? I grew up in rural southwestern Ohio, a little under an hour's drive from downtown Cincinnati. When I was six, we moved to a 12-acre (~5-ha) portion of an 80-acre (32-ha) farm subdivided for housing. The immediate landscape had almost no trees: it had been plowed farmland before being broken up for housing. On one side, our property was bordered by a cattle pasture with scattered scrubby and thorny trees and a wooded ravine. At the far end, however, it bordered an honest-to-goodness forest, which inexorably called to me and was separated only by the scantest remains of an old, barbed-wire fence.

When I was small, I had to stay close to the house unless my father walked with me. On weekends, when my father was not too busy (and he always made time), we would walk through the "high stuff," the weeds that sprung up in the large portion of the acreage we did not mow. Sometimes, we would leave our property, cross the fence, and walk along the creek in the forest—these were the best times. As we would walk, my father noticed things. I remember him pointing out a brown-headed cowbird (*Molothrus ater*) egg in another bird's nest. Cowbirds are brood parasites, laying eggs in other species' nests so that they get their young cared for, often to the detriment of the host bird's own young. He lamented the loss of purple martins (*Progne subis*), insect-eating birds that nest in colonies. My paternal grandfather built martin nest boxes and put them up on poles at just the right time to get martins but not the dreaded European starling (*Sturnus vulgaris*, who needed no help anyway). My father's childhood memories included watching these elegant birds dart and forage on flying insects in the evening sky.

Occasionally my father would tell stories, often abruptly with little to no warning. One of these childhood stories concerned his relationship with trees. Growing up, my father and his siblings would collect black walnuts (*Juglans nigra*) from the trees in their back yard, husk them, and crack them for the nuts. The outer covering on the walnut smells strongly of chemicals; I have no words to describe it other than it smells like a walnut. As this husk breaks down, it produces a dark brown pigment—the original walnut stain. When you remove the husk, your hands become colored with it, and it leaves a deep stain. My father and all his siblings went to the local Catholic school, and the nun instructed my father to wash his hands when he appeared all brown-handed after processing walnuts. "Sister, it won't come off," he replied. Sent to the bathroom, he returned, showed his hands, and repeated, "Sister, it won't come off." Unbelieving, she took him to the bathroom and scrubbed his hands. When they remained gloriously soiled, she marched him to the adjacent priest's house where all available cleansers were brought to weigh on the issue, to no avail.

Decades later, my grandfather still husked the walnuts himself. He would disappear into the old chicken coop after dinner and present us, brown-handed, with a sack of black walnuts to take home. He would dig the walnut seedlings that came up in his vegetable garden each fall and give them out. The first trees at my childhood home were those black walnuts, planted in a row along the back fence. I have seedlings from those trees planted on my property now.

As I got older, I could roam farther on my own. In agricultural landscapes, the ravines and other nontillable areas often remain as forests for wood production, hunting, and maybe grazing. The little strip of woods adjacent to our land was connected to an extensive network that followed the creeks and streams for miles and miles. That network of riparian forest became my highway to freedom. I, however, was not the only thing moving in that landscape. Over the years, I watched the forest cross the fence onto our property—first a few scattered trees, heavy-seeded oaks (*Quercus*) closer to the fence line, wind- and bird-dispersed ash *(Fraxinus)*, hawthorn (*Crataegus*), and cherries (*Prunus*) colonized farther out.

Then, forest understory plants began to cross—mayapple (*Podophyllym peltatum*) the most obvious, spreading clonally a few feet farther each year. Year after year, foot by foot, our property became more forested, the old field goldenrods (*Solidago* spp.) and grasses yielding before the progressively taller shade cast by woody plants.

I followed my love of trees and forests to college, where I majored in botany at Miami University (in Ohio; a university before Florida was a state!), continued into a master's degree, and then on to a doctorate in ecology at Rutgers in New Jersey. I spent much of my 6 years in New Jersey at the Hutcheson Memorial Forest, conducting my dissertation research, contracting Lyme disease, and leading public tours. During this time, I also began working on a long-term study of vegetation recovery after agriculture, a project to which I would devote much of my professional life. I continue to return to New Jersey annually and have shepherded that succession study into its sixth decade. My long experience with Hutcheson Memorial Forest and the old-growth forest that serves at its core brought me to reflect on forests and their dynamics over time, their past, their present, and looking toward their future. This thought process ultimately led me to work focused on the trees that I love, individually and collectively, and on the threats that trees and the forests face.

## Goals for This Book

After years of leading tours and lecturing on trees and forests and the threats to them, I realized that when people hear about forest problems, the problems are categorized mentally as separate facts rather than interconnected and compounding threats. As independent, unconnected incidents, the gravity of the situation appears lessened. In part, this separation is generated by the necessary regional aspect of forests—at any one time, your local forest is likely facing only one or a few challenges, perhaps with recovery between each event. This perspective is also facilitated by viewing forests as first and foremost a resource, where the loss of a tree species is a temporary loss to the tree canopy. If society more fully appreciated how forests function, the ecological roles of individual species in them would be a more prominent part of the conversation.

This book will help to make those connections clear in a straightforward manner. My aims are relatively simple—I want to present the issues facing some of our major tree species and why, perhaps, you should care. This book intends to offer a scientific view rather than a historical, sociological, or political perspective. Other books cover those perspectives. I will try to keep the text as nontechnical as possible—explaining the necessary principles as we go without being unnecessarily encyclopedic.

I hope that contributing to the discussion of Eastern North America's forests will help lead to more informed decisions, regardless of whether society collectively chooses to face the problems or chooses to ignore them. We should at least appreciate what we risk before making this choice. It might come as a surprise that I am optimistic about the future of our forests. Species invasions can and have been managed or even removed entirely. Forest managers can do much to steward healthy forests for generations to come. The key is to ensure that we halt the additional challenges that we are heaping onto our forests. The ability to safeguard our future forests' health is dependent on allocating resources toward their defense and management. Resource allocation, in particular financial resources, get apportioned only to things that we inherently value. My childhood was blessed by a myriad of interactions with trees and forests. I would desperately love for future generations to have similar opportunities and to grow to value those same forests. And that is the purpose of this book.

# Acknowledgments

First, many thanks to friend and colleague Mary Cadenasso. Your response to an offhanded comment on a stroll in Sacramento—"That would be a great book title . . . you should write that book!"—started me on this journey. I do not think it would have occurred to me otherwise. My thanks to everyone at Cornell University Press, particularly Kitty Liu, the editorial director of Comstock Publishing Associates, for all the help and guidance along the way. Your immediate enthusiasm for the project meant more than you can know. This text also benefitted greatly from comments on early drafts from three former students: Kirstin Duffin, Anna Herzberger-Chen, and Kelsey Phipps-Lessaris. It was fun to reverse the direction of editorial comments for a change. I also owe a great debt to Roger B. Beck, friend, neighbor, and copyeditor. You were tough and thorough, and I ignored only a few of your comments. Finally, and most of all, thanks to my wife Melissa, who has (mostly) patiently listened to me prattle on about these sorts of things for decades now. You are the best!

# TREE BY TREE

# INTRODUCTION

## *First, Some Context*

What do I mean by the decline of forests? Will all the forests be lost, leaving us with a landscape barren of trees? Well, no—although global climate change may indeed result in the loss of forests in arid areas such as the Sierra Nevada of southern California. Fire frequency is predicted to increase, making the time between burns too short to allow trees to persist in that region, whereby they will be replaced by shrubs better adapted to fires. Changes in tree density or composition are expected throughout many areas in western North America as fire frequencies and intensities increase. However, there is nothing that suggests Eastern North America will lose tree dominance over large areas. What I mean by decline is more in the metaphorical sense—the loss of the way the forests were historically.

Think of the reasoning that a historic preservation committee may use in defending a treasured neighborhood or downtown area. The loss of the town square's traditional businesses, diners, and small shops and their replacement with title agencies and lawyers' offices alters how people engage with, utilize, and interact in that space. Such changes result in the decline of a historically vibrant downtown community, particularly after the law offices close for the evening. Similarly, the conversion of a historical neighborhood from stately old homes, perhaps in need of love, to strip malls and apartment buildings can lead to neighborhood decline. There are still buildings and streets, but the feel of the place changes dramatically. The diversity of building types, stores, and services that characterized the historical community is lost. The area's function has changed—some will see this as a positive as tax revenues may certainly increase; some will see this as a negative; some will not care because they do not live there; and others will care because they do.

The types of changes to forests addressed in this book are along similar lines. Mainly through human activity, several major tree species in our forests have gone through, or are facing, massive declines. As in the conversion of a historical neighborhood, when one business closes, or an old home is too decrepit to restore, something replaces it. The same general rules govern the processes functioning in both the economic and ecological scenarios outlined—they both come from the same Greek root after all—*oikos*, meaning home. In real estate, a vacant property represents a loss of income from business, rent, or taxes. Quite simply, there is no reason to hold onto an unused property. The loss of one building is typically an economic opportunity to build a new one. In forests, when a tree species suffers a massive die-off, such as the American chestnut (*Castanea dentata*) due to chestnut blight, there is an ecological opportunity for another species to move in to fill the openings. Even smaller-scale disturbances allow for the growth or introduction of new plants into a forest community.

The analogy between neighborhoods and forests ultimately fails because forests are much more complicated than towns and have a much deeper history. As forests are inherently dynamic features of the landscape, picking a historic point in time with which to compare modern forests is difficult. Are the forests of your childhood the target? Or perhaps the forests before European colonization or maybe even the ones before indigenous peoples colonized the area. Your perspective ultimately determines your view of forests, how you interact with forests, and the value you place upon them.

In forests, a primary driver of change is not economic pressure but rather the availability of soil resources and solar energy that remain unutilized. When an individual tree dies within a local forest, something replaces it. If environmental conditions still favor trees over other plant life-forms, a tree should come to dominate that spot. If the plants that colonize the opening initially are not trees, these early colonizing plants will eventually be displaced by trees through microscale successional processes (the regeneration of forests following a disturbance). If we scale this recovery up from the loss of an individual tree to the loss of an entire species from a local forest, we see a similar process operating at a much larger physical scale.

Another critical driver of change in forests is the rate at which trees die, representing opportunities for regeneration. These rates also change in space and time. Storms generate blowdowns or other canopy disturbances, insects and pathogens attack individual trees, and droughts weaken and kill trees overtopped by larger individuals. Species tolerant of shaded conditions dominate forests with low mortality rates and fewer canopy opportunities. Increase the rate of tree death, and you will shift the system toward tree species better able to deal with canopy openings, altering forest composition. As tree species differ markedly in their resistance to droughts, fires, and pests, we expect more significant effects on some species than others. Climate change, with the accompanying weather alterations, and the introduction of new tree-specializing insects and pathogens have increased mortality rates for many canopy species based on their susceptibility to those threats.

Forests or other landscapes with trees will persist as long as conditions allow trees to regenerate. Forest composition will change with some tree species expanding to fill the ecological vacuum, the unexploited resources that remain when a tree species is lost. The forest that will remain will be less diverse and, likely, will function differently than the historical forest, but it will be a forest, nonetheless. Whether we value the new forest will depend on our personal views, how we interact with that forest, how we value the individual species, and whether we even notice the change.

I could say that Eastern North America's forests are currently facing an unprecedented threat to their composition—but that is not exactly true. We have precedents, which we have repeatedly ignored. This book will explore two tree species already functionally lost from our forest communities—the American chestnut and the American elm (*Ulmus americana*). As the events that threatened these species mostly occurred before many of us were born, we—myself included—have little direct experience with these trees. The remaining species to be discussed are those currently experiencing massive threats that may ultimately follow a similar path toward ecological novelty rather than persisting as dominant forest species. The question remains—with the loss of major tree species, why hasn't there been more of an outcry?

I can summarize my theory for the lack of public notice of the forest's plight as "the world is green, green is good." This idea first came to me when I realized that much of the vegetation that most people interact with is, in fact, not native to North America. Nothing clarified this more for me than when a local New Jersey phonebook had a picture of a glorious stand of purple loosestrife (*Lythrum salicaria*) emblazoned on its cover. This admittedly gorgeous Eurasian plant is wildly and widely invasive in freshwater wetlands, often dominating invaded wetlands. Why do people not notice when species from far, far away invade their forests and wetlands? It is because we tend to look at our landscapes holistically. As long as our landscape remains green and vegetated, maybe with some lovely flowers or delicious fruit, all is good, regardless of the origin of the plants. If something were to happen so that the landscape was not so green, we would notice in an instant, and an outcry would be raised—at least I want to believe it would. This unawareness of plants has been called "plant blindness" by some researchers and represents a major forest conservation challenge.

Many non-native plants have persisted in local forests for more than a human generation, so even our personal experiences are not reliable ways to assess our environment. Childhood daisy chains are made predominately with an early colonist (*Leucanthemum vulgare*), a Eurasian plant that was brought over by early European settlers. The honeysuckle of my youth that you could sip nectar from is an Asian species (*Lonicera japonica*) introduced for its fragrant blossoms and rapid growth, resulting in success for even the worst backyard gardeners. The continually shifting frame of human perception is a significant barrier to understanding the plight of our forests.

The natural resiliency of forests is perhaps why the loss of individual tree species has not been noticed on a scale proportional to the area of the landscape that has been affected. Our forests have a long history of resource (i.e., tree) extraction, so we are used to forests regrowing and changing over time. Loggers had already cut much of the Eastern forest at least once by the time the American chestnut (chapter 2) vanished from the landscape. The ecological opportunity left by the loss of this species was filled in by oaks (*Quercus* spp.) and hickories (*Carya* spp.) as the forests regenerated following clear-cutting. Similarly, the loss of American elm (chapter 1) from wet forests led to increases in other tree species. Largely undocumented, associated species dependent on those tree species for their existence were also lost. We did not lose forests as an entity of our landscape, but we lost a shade of green; we lost diversity. However, what remained was still forest, still green, still—by default—good. Hopefully, this book will change that perspective so that individual tree species, not just forests, become a more integral part of our daily discussions and cares.

We, as a society, are now a much more urban population than ever before. Fewer people are directly interacting with forests. Urban trees are typically relegated to small plantings along roadways and parks and grow as the weeds of the brownfields that are all too common in cities. Despite the broad benefits of trees in cities, they tend to be primarily in wealthier neighborhoods, leading to environmental injustice based on socioeconomic and racial factors. Suburban populations often have more significant interactions with trees but less so with forests. As cities expand outward into the surrounding landscape, they often convert former agricultural land into housing subdivisions. Pre-existing trees are typically removed and replaced with fast-growing ornamentals that would otherwise not exist in this landscape. While these replacement trees provide the ecological services of cooling shade, nutrient retention, and reduced air pollution, these plantings are by no means functioning forests. With fewer and fewer people in successive generations experiencing tangible and significant interactions with natural forests, we should not be surprised at our ignorance of the threats to them.

## The Focus on Eastern North America

A text focusing on forest tree problems such as those discussed here could include species from many areas of the planet, perhaps identifying trees from a variety of forest types to exemplify the problems' global nature. However, that scale of focus runs the risk of presenting the issues as isolated and peculiar to each tree species—case studies without a single, unifying location as a context. Selecting a focal region highlights not only the details of the challenges to individual tree species but, in aggregate, represents an honest and at least partially complete view of the threats to regional forests. Eastern North America has suffered the large-scale loss of not one but two common trees and likely faces more in the coming decades. For this reason, I have selected to concentrate on the forests of this area, which is, conveniently, also where I have lived and worked my entire life. I have witnessed many forest changes and their continuing impacts. Some I have witnessed directly; of others I have seen the aftermath as persistent ripples on the landscape.

What other reasons make this geographic area a cohesive focus for a book? This area represents the New World region with the longest history of intensive European colonization, habitat conversion, and forest management, or lack thereof. Most of the region's forests have been cut at least once, many two or more times since European colonization. Conversion of primary forest to agriculture or cutting for timber has already occurred on a large spatial scale resulting in old-growth forests being a rarity in Eastern North America. Many forests have since regenerated and are now more or less persistent in the landscape. The period of resource extraction and forest conversion occurring in many tropical systems today is something that Eastern North America has already completed. The long history of European occupation in Eastern North America is an equally long history for species introductions. Early on, European settlers moved plants and animals, purposely and accidentally, to this continent, forever changing the region's species pool. The sustained economic activity of Eastern North America over the centuries has ensured the continued movement of species, including plants and, more importantly, their pests, into the region's forests. It is this economic activity that has primarily resulted in the threats upon which this book will focus.

My observations on human activity's role in the region are backed by two studies generated by the National Center for Ecological Analysis and Synthesis (NCEAS). Using large databases of species records, a group of researchers examined forest pests' temporal and spatial patterns in the continental United States, including pathogens and plant-feeding insects. Their results are more disturbing than one could imagine. They found 450 species of insect pests and 16 species of plant pathogens that have become established since European colonization. Most insects have not become extensively damaging, but 14% of these species have become severe pests that generate economic impacts: that is sixty-three species of insect pests that damage forests. Even more alarming, all the pathogens have had harmful consequences. These numbers are likely vast underestimates, as we tend to notice the more damaging pathogens and insects that appear in the landscape. Less detrimental insects, particularly those from less noticeable and noncharismatic groups, probably lurk undetected. Everyone loves and notices beetles; almost no one loves microlepidopterans—tiny moths. Furthermore, fungi and bacteria are all but invisible unless you know to look and care to do so.

Human movements, which create routes of introduction, have resulted in the invasions of forest pests that are heavily clustered in the northeastern states. Invasions thin out in the central states, much as the density of forests does. There continues to be a second, less-organized wave of invasion moving inland from the West Coast, a second center for people and material movements. The NCEAS research estimates that the United States accumulates an average of slightly more than 2.5 forest insect pest species a year. This rate has remained remarkably unchanged for the last 150 years. This time includes periods before the Plant Quarantine Act of 1912, the first federal import regulations to control the human distribution of plants. Since then, there has been continual development of rules and regulations on importing plants and plant materials, culminating in the modern incarnation of APHIS (Animal and Plant Health Inspection Service) today. All this work has resulted in holding the rate of forest pest introductions constant, but it has not reduced them. This level of success would seem to be an abject failure if not for the dramatically increased flow of materials and plants that now require monitoring.

Once in North America, these insects and pathogens move throughout the forested landscape, interacting with native species as they spread. One NCEAS paper estimated that these forest pests move 3.2 miles (5 km) each year during their invasions. The actual spread is much more erratic, with periods of relative stasis because of habitat fragmentation or distances between populations of suitable host trees. Slow periods of movement are offset by long-distance leaps into new habitats, followed by radiation outward from those new populations. Many forest pests hitch rides with the constant movement of people from place to place. Sadly, this includes even nature lovers' movements when they accidentally transport pests via vehicles, gear, and the mud on their boots, resulting in targeted dispersal from habitat to habitat.

The average movement rate of forest pests seems relatively slow by human standards. However, if you think about this movement from a management perspective, it is horrifying. Assuming a simple spread out from a point, one year's expansion of a typical forest pest would create more than 32 square miles (21,000 acres; ~8500 ha) of potentially infected forest trees. If we use a middle-of-the-road forest density of 120 trees per acre, we find an astounding 2,552,000 trees in that area may be infected. Of course, the tree that is the potential host for the new invader would represent a subset of the total density. Someone would still need to inspect, treat, or destroy every individual tree in the potentially infected area to control a forest pest successfully.

Despite the long history of human alteration and use, Eastern North America's forested landscape is surprisingly large and continuous. From Canada's boreal forests to the pine-dominated woodlands of the southeastern United States, trees are the potentially dominant vegetation throughout. These forests, though changing in composition, extend westward until the decreasing moisture and historical fire regimes replace tree-dominated habitats with the central prairie region's grass-dominated habitats. Modern agriculture has displaced much of the original forest vegetation, but forests remain in those areas unsuitable for agriculture or otherwise set aside. There have also been areas of large-scale forest recovery as agriculture has shifted from the less productive soils of New England to the continent's central corn and wheat belts. The pre-European area of Eastern North America's forests represented a massive, unbroken area of forest. Folklore suggests that squirrels could travel for hundreds of miles without touching the ground; similarly, insects, fungi, and plants found an equally continuous habitat for movement across the landscape.

Contrast this mostly continuous block of the Eastern forest with the forests of the western United States. Heading eastward from the coast of California, you find the Coast Range (forested), the central valley (forests restricted to riparian habitats), the Sierras (forested), the Great Basin (unforested), and then the Rocky Mountains (forested). While each of these regions is large, they are separated physically, environmentally, and compositionally. Many western tree species are isolated to one of these mountain ranges or have discontinuous distributions determined by altitude. The dramatic changes in the environment that separate these habitats strongly limit species' ability to spread across the landscape. The much shorter span of altitudes in the eastern mountains represents a lower barrier to species' movements than in the mountains of western forests. Overall, tree and other species are much more continuously distributed in Eastern North America than in the rest of the continent—for both good and ill. The broad distributions result in trees that may be dominant in some forests, common in others, and occasional in still more. Therefore, when a tree species is lost, the impact can be felt throughout an extensive area.

Similar tragedies occur in other habitats around the world with the same potential level of damage. Eastern North America is by no means unique. Sudden oak death and population explosions of pine bark beetles threaten their host species across a broad region of western North America. North America itself is not unique in the threat to forest tree species. Dutch elm disease, the disease that wiped out the American elm (the focus of the next chapter), also occurs in Europe with similar effect on elm populations. Ash (*Fraxinus*) trees are threatened in Europe by both a fungal disease (ash dieback disease, *Hymenoscyphus fraxineus*) and an outbreak of the invasive emerald ash borer beetle (chapter 4), with potentially catastrophic effects. Similarly, myrtle rust (*Austropuccinia psidii*) was introduced into Australia from South America and threatens many tree species in the Myrtle family, including *Eucalyptus* species. Climate change threatens tree species everywhere, but the effects of that will likely be a slower process than the tragedies that we will focus on in this book.

Most forest threats in Eastern North America have occurred because of the spread of pest species from the temperate Old World to the temperate New World (Eastern to Western Hemispheres). The reason is a combination of the climatic similarity and the biogeographic affinities between the regions, which will be the focus of the next section. The net movement of species has driven this transfer's directionality from the Old World to the new. In fact, the New World flora was considered inferior to that of the Old World by early botanists, who advocated for the transfer of "superior" European species. While that opinion has diminished, the U.S. government often played a role in species introductions across the region. While not responsible for the original introduction, the Soil Conservation Service spread wonderfully invasive species such as *Rosa multiflora*, promoting it as a living fence. Even after the rose's ability to take over pastures became apparent, it was still encouraged as cover for wildlife and food for birds, who spread it far and wide from those original plantings.

Of course, some examples reverse the trend of species movement from the Old World to the new. Most famously, phylloxera (*Daktulosphaira vitifoliae*), an aphid-like insect that feeds on North American grape (*Vitis*) species, was accidentally introduced to Europe along with North American grapes. The insects tagged along with their grape hosts, but the European grape lineages had no evolutionary exposure to the pathogen, resulting in the widespread collapse of wine production in the late 1800s. Growers eventually brought the disease under control by grafting European wine grape stems onto rootstocks from North American grape species, which retain resistance to the pest.

Non-native plants from Eastern North America have also done quite well in Europe for the same reasons European species have done well in Eastern North America—similarity of climate. While riding on a train in Europe, you can watch the landscape roll by, filled with the same species you would see traveling by train in Eastern North America. You can see goldenrods (*Solidago* spp.), common milkweed (*Asclepias syriaca*), black locust trees (*Robinia pseudoacacia*), box elder (*Acer negundo*), and many, many others. You get similar importance of New World plant invaders in Asia's temperate locations as well, though they are fewer in number so far. Ragweed (*Ambrosia*) species, the bane of hay fever sufferers, has become a successful weed in many Asian habitats, as have the showier goldenrods often unfairly blamed for the allergies.

As species move among the world's tropical and subtropical regions, we should expect similar invasion problems to develop because of their climate similarity. Hawaii and Florida are famous for their mild climates that allow for the development of lush, tropical gardens—resulting in their floras exploding with non-native plants and animals sampled from across the world's tropical climates. Pests of tropical crops have colonized with similar effects as those on native tree species. Of much interest to Florida is the colonization of orange orchards by the Mediterranean fruit fly (*Ceratitis capitata*). This African insect has been unsuccessful to date but incurs massive costs annually to prevent its colonization. A successful orange pest to colonize Florida is the Asian citrus psyllid (*Diaphorina citri*; a type of plant louse) and the citrus greening disease (*Candidatus liberibacter*, a bacterial pathogen), which the psyllid spreads. As global commerce increases, and the world's range of crops becomes more uniform, such pest and pathogen movements will likely become more common. Customs forms ask about fruit in your possession for a good reason.

I present the challenges to Eastern North America's forests as a case study, a warning for the rest of the planet. While Eastern North America has arguably been hit the hardest, there is nothing special about the tree species here that will prevent similar things from happening elsewhere. Evidence suggests that similar species losses are sadly happening in both natural and agricultural settings in a great many places worldwide.

## The Changing Context of North American Forests

The first documented functional extinction of a tree species in North America occurred early in the continent's post-European colonization history. In 1765, John Bartram, the royally appointed botanist, started an expedition to the southeastern United States, collecting plants as he went. He discovered a small flowering tree with fragrant white blossoms along the Altamaha River in what would later be southeastern Georgia. His son returned to that location in 1773 to collect seeds to grow in the Bartram's garden. The senior Bartram named this tree *Franklinia alatamaha* after his contemporary, Benjamin Franklin, and the river of its origin. After a few early sightings, the tree was never seen in the wild again, persisting now only in botanical gardens and gardening aficionados' backyards. This story of discovery and loss is repeated time after time in gardening circles. Of course, the species is not extinct, propagated in perpetuity, but it has disappeared from the native plant communities in which it occurred. The Bartrams found the plant at the end of its natural trajectory. As the loss predates the wide-scale alteration of the landscape and environment by agriculture, industry, or other activities, humans cannot be blamed. This loss was a natural extinction event.

Forests—all plant communities really—are always changing. How can we place the present losses of dominant canopy trees into the context of an ever-changing ecological landscape? This is the challenge. Surprisingly, the dynamic nature of plant communities is a relatively new perspective on forests. Significant forces that drive forests, such as fires, only began to appear in ecology textbooks in the 1970s. Early ecologists treated communities as relatively unchanging. When a forest was cut down or otherwise destroyed, it would recover through the process of succession back to its original state—often referred to as the climax community. The forest transitions would be from shade-intolerant, sun-loving trees toward shade-tolerant species that can regenerate in a closed canopy forest. The belief of ecologists who followed this view was that forest dynamics occurred only during forest recovery, and once the original composition recovered, the community would be stable at its climax. Since then, the climax community concept has become a persistently dirty word in ecological circles. Scientists who went through the conceptual transition away from the climax concept remember the long arguments and entrenched philosophies that impeded progress.

What eventually replaced this early restrictive view of dynamics was a focus on individual species' tolerances. Under this perspective, a tree species would grow wherever its seeds dispersed if it could persist under local environmental conditions. If conditions changed, the species composition would also change. This dynamic view is a far cry from the stability suggested by the idea of a climax forest. A period of drier conditions would move forest composition toward species that did better under those conditions. A population increase in an herbivore, such as deer, would change composition toward species tolerant of or resistant to herbivore damage. Trees falling during storms would allow shade-intolerant species to persist in a forest dominated by shade-tolerant trees. A period of increased storm activity would result in a pulse of shade-intolerant tree regeneration that would persist in the forest canopy for as long as those trees lived. Modern ecological thought embraces the continually changing nature of forests and attempts to incorporate this philosophy into forest management.

Interpreting the contemporary losses of forest trees against a backdrop of dynamic forests is difficult. Climatic fluctuations, increased herbivore pressure from white-tailed deer (*Odocoileus virginianus*), and increased pressures from resource extraction affect nearly all of Eastern North America's forests. However, once we build an appropriate context for the forests, interpreting tree losses becomes a more straightforward task.

## Global Floristic Patterns

To paint an accurate picture of how the forests of Eastern North America have come together and how they relate to other areas of the planet, we must first examine the development of North America's flora deep in time as the continents have assembled. Two things link the floras and faunas of the Northern Hemisphere—their physical connection in the not-too-distant past and the fact that they have remained primarily in a temperate climate over their history. When Pangea, the single large mass of land, broke up approximately 200 million years ago, it formed Laurasia (North America, Europe, and continental Asia) and Gondwanaland (everything in the Southern Hemisphere, plus India). The Nearctic (North America) and Palearctic (Europe and most of Asia) generally stayed together in the Northern Hemisphere, with connections appearing and disappearing with sea level changes. Northern Hemisphere continents were connected most recently 50 million years ago. The continued similarity of climates into the modern era means that species from one latitude in the Old World are already more or less preadapted to that temperature regime at a similar latitude in the New World and vice versa. Proximity to mountain ranges and moisture sources determine local rainfall, but large sections of North America, Europe, and Asia are temperate with moderate rainfall amounts, referred to as mesic habitats. The overall similarity of climates across the Northern Hemisphere has resulted in the ability of insects, mollusks, mammals, birds, and fungi to move among continents, establish persistent populations, and occasionally cause problems.

Floristically, temperate regions of the Northern Hemisphere are the dominion of the pine and its relatives. Despite the evolutionary advances of the flowering plants, the gymnosperms (conifers) dominate. There are pines (*Pinus*), spruces (*Picea*), firs (*Abies*), and junipers (*Juniperus*) found in North America, Europe, and Asia. Coniferous forests often have fewer species, with only one or two dominant species. If we restrict ourselves to the deciduous forests of temperate latitudes, we find forests structured similarly on all three continents. Deciduous trees dominate the forest canopy, with lianas (woody vines) often linking the forest canopy to the forest floor. Under the canopy layer is a subcanopy composed of small trees and shrubby species, and below that, an herbaceous layer. The herbaceous flora contains two types of plants: spring ephemerals that largely complete their life cycle and go dormant before the forest canopy fills in each spring and shade-tolerant plants that persist and are photosynthetically active throughout the growing season.

Taxonomically, there is a lot of overlap across the Northern Hemisphere as well. Oaks occur on all three continents. The North American

native red oak *(Quercus rubra)* has become established in Europe and is
invasive in some areas. Sawtooth oak (*Q. acutissima*) is native to China
but locally invasive in the United States. On the forest floor there are gin-
gers (*Asarum*) and jack-in-the-pulpits (*Arisaema*). A backyard gardener in
Asia, Europe, or North America can find *Viburnum* shrubs at their local
garden center that are native to another continent of the Northern Hemisphere and grow them successfully on their home continent. *Aster* species
are common throughout, though taxonomists have now separated the
New World species into a separate genus, *Symphyotrichum*. Many additional tree and shrub genera occur in both North America and Eurasia,
such as Alder (*Alnus*), basswood (*Tilia*), beech (*Fagus*), birch (*Betula*),
dogwood (*Cornus*), hickory (*Carya*), horse chestnut or buckeye (*Aesculus*), poplar (*Populus*), and willow (*Salix*)—just to name a few. If you go
back further in time, ginkgo (*Ginkgo biloba*) and dawn redwood (*Metase-
quoia glyptostroboides*) are found as fossils in North America and Europe
and could be considered native to both continents. However, their current
natural distributions lie wholly within Asia. If we added lianas, shrubs,
and herbs, the shared list would become even longer. When these plants
are introduced into local forests or as fun new additions to your backyard
garden, they become potential new invasive species for your region and
neighborhood.

More important to this book's purpose is the geographic distribution
of five key genera, representing the canopy trees that have been lost or face
becoming lost from Eastern North America's forests. The elm (*Ulmus*),
the subject of the next chapter, occurs in North America (*U. americana,
U. rubra*), Europe (*U. laevis*), and Asia (*U. parvifolia*, invasive in North
America). Chestnut's distribution is similar, with North American (*Castanea dentata*), European (*C. sativa*), and Asian species (*C. mollisima*,
commercially grown in North America). The maples are a very species-
rich group with several species in North America (in the east, chiefly *Acer
rubrum* and *A. saccharum*), Europe (*A. platanoides*, invasive in North
America), and Asia (*A. sieboldianum, A. palmatum*, and *A. japonicum*, to
name a few of the many grown horticulturally). The ash tree also occurs in
all three continents represented by several North American species (*Fraxinus americana* and *F. pennsylvanica*, the most widespread), European
(*F. excelsior*), and Asian species (*F. mandshurica*). Last, hemlock occurs
on both the east (*Tsuga canadensis*) and west (*T. heterophylla*) coasts of

North America, as well as eastern Asia (*T. chinensis* and *T. sieboldii*), but is absent from Europe. This list is not exhaustive, but it begins to illustrate how many of these tree genera are shared across continents.

This genus matching across continents is not just an odd coincidence but rather represents an ecological challenge when species move between distant areas. These disparate areas share species with similar evolutionary origins and similar current environmental conditions, allowing many species to successfully grow and form persistent populations outside their historical distributions. However, the time that the continents have been separated is time that evolution has not stood idle. Natural selection has continued to act upon the populations of the now-isolated species as they have contacted herbivores, pathogens, and other challenges to their populations. As new enemies have evolved, selection has favored individual trees with genes that have allowed them to defend or survive the new threat. Tree responses to enemies, in turn, have acted as a selective pressure on the pathogen and the herbivore, favoring genotypes that can deal with the plant's new defenses. Cycle after cycle, host and enemy populations have evolved reciprocally.

However, as plate tectonics have resulted in the isolation from and loss of former pathogens or herbivores, selection still functions in another direction. Plant defenses often involve the costly use of carbohydrates (energy) and mineral nutrients to ward off insect pests or diseases, whereas these nutrients would otherwise be allocated to a plant's growth and other purposes. A plant must be conservative in its defenses. The nutritional and energetic costs associated with protecting against a nonexistent natural enemy reduces the resources it can allocate to reproduction and fitness. Natural selection has favored plants that maintain a balance of resource allocation between defense, reproduction, and growth. Without continued benefit from defense, a plant's ability to defend is reduced, potentially causing it to lose—at the extreme end of the spectrum—all ability to protect itself from damaging agents. Consider the ill-fated dodo that became flightless and predator foolhardy in its long isolation from any predatory animals until humans arrived. At some point, the flightless dodos had more offspring than those that retained the ability to fly and were favored by selection. That is an oversimplification of the process, but I think the point is clear. Species continually change in response to the local conditions that they experience.

So, where does this leave the forests of Eastern North America? Similar enough to the floras and habitats of Europe and eastern Asia that species from these areas can invade North American forests; likewise, North American plants can become invasive if transported to the Old World. Dissimilar enough from species in Europe and eastern Asia that North American species may no longer have resistance to, or may never have evolved resistance to, Eurasian herbivores and pathogens, leaving them susceptible to pest and pathogen outbreaks.

While the focus of this book is on the challenges associated with introductions, these interactions need not lead to problems for North American species. One of the most successful non-native plant invasions is multifloral rose (*Rosa multiflora*), introduced into Eastern North America from Asia, mentioned earlier in this chapter. This rose was actively introduced as a living fence and spread slowly westward. Eventually, the invader encountered a native western rose species that had a local viral pathogen. The Asian rose species had never been exposed to this North American disease and had little resistance. Rose rosette disease has since swept eastward, decimating the invasive rose's populations as it moved from plant to plant by native sap-feeding insects. It is marvelous to wonder at how many plant species would otherwise become widely invasive but never get the opportunity because they succumb to a local herbivore or disease before we even notice them. Such unseen interactions are certainly food for thought.

## Temporal Dynamics of North American Forests

While the floristic patterns of plate tectonics manifest themselves over deep time, there are equally important forces operating within North America that have more recently shaped our forests. It is important to place the recent loss of tree species against an appropriately dynamic backdrop as a context. Tree populations have always responded to climatic changes, generating changing forest composition over time. The changes in forests discussed here will focus on dynamics occurring on a time scale of hundreds to thousands of years.

Across the Northern Hemisphere, glaciation has been a significant force in structuring the landscape and the flora that occur there. Over the millennia, wave after wave of ice sheets thick enough to deform the planet's crust have pushed southward then receded. The last major glacial period in North America was the Wisconsin glaciation, starting 71 thousand years ago and ending a mere 12 thousand years ago. We tend to envision glaciers as single entities that sweep out of the north, reach their maximum, then recede poleward in an orderly manner. Nothing could be further from the truth as glaciers are quite dynamic. The weight and pressure from the glacier's main body can push ice forward like a fluid, physically shoving the front edge forward during ice accumulation periods. This advance may be followed by a hundred years of recession, followed by another rapid advancement period. The dynamism expressed in glaciers is not of much note in geologic time, but it can be crucial in developing local landforms and to the organisms that live there.

From a landform perspective, the most recently glaciated terrain is poorly drained, as 10,000 or so years is not enough time to develop the uplands, lowlands, river corridors, and other features necessary to facilitate water movement. Such poorly drained yet rich soils are the dominant landform in the upper and central midwestern states and Canadian provinces and are currently employed in row crop agriculture, now drained by cultivation practices. If we move far enough northward, the boreal forests in Canada are essentially forested wetlands. Mountain ranges deflect the advancing glaciers, though they also succumb to the massive ice sheets' grinding action and show glacial scouring in many locations. John Muir boldly hypothesized glaciers' role in forming landscapes in the Sierra Nevada mountains and was ultimately found to be correct. Some glaciers penetrated farther south; some did not. The key for any location is less about how many times an area has been glaciated but, rather, how long since the last glaciation event. Of course, the advancing ice misses some land areas. Famously, Appalachian coves and the driftless region of Wisconsin missed much of the ice and remained as islands of diversity where plants and animals waited out the long glacial period to recolonize when the ice receded from around them. All other organisms were displaced farther and farther south with the advancing ice and cold temperatures. Those species that could not move quickly enough were lost.

As the glaciers moved southward during a building phase, they pushed up debris—soil, rocks, and the occasional tree at the leading edge. Much of this material did not get displaced far, but occasionally things moved great distances. Much of middle North America's geology is sedimentary: sandstones and limestones laid down when shallow seas dominated the area. The farm fields on glaciated terrain throughout this region often contain nonsedimentary granitic boulders, transported massive distances from their Canadian origins. Most of these are small and rounded, ground smooth on their long glacial ride, much like you would see in a riverbed. Occasionally you will find larger stones, some as large as Volkswagens, placed at the borders of a farm field. When the glaciers finally receded, this material was left behind, remaining as a long ridge of everything that the glacier has picked up, leaving what is called a moraine on the landscape.

During periods of glacial expansion, species moved southward according to their cold tolerances. When the glaciers began to retreat, vegetation soon followed. What was left behind by the glacier was not so much soil as a pile of minerals—there was no organic matter or organisms among the bits and pieces of rock to make it soil. Mosses and algae first colonized the newly exposed substrate, followed by small herbs, often those that can pull nitrogen from the atmosphere to make up for the soil's lack of nitrogen. Over time, their root and shoot tissues decomposed to build up the soil's structure and fertility. Next, shrubs and small trees advanced, then coniferous trees—boreal spruces (*Picea*) at first, followed by some hemlock (*Tsuga*). Forest development may take 200 years or so if uninterrupted. Glacial retreat can be just as dynamic as a glacial advance and can quickly reverse directions. Glacial advances can reset the vegetation and may even bring a new glacial deposit on top of the old vegetation, resulting in some interesting fossil remains. In a quarry in central Illinois, close to where I now reside, there are spruce logs buried under a cap of clay soil, preserved by the anaerobic and dense layer of clay dropped on them more than 20,000 years ago during the glaciers' last hurrah. Spruces are a recognizable vestige of a plant community that no longer resides in the state but now dominates hundreds of miles northward into Canada.

As the glaciers continued to retreat and the climate continued to warm, the forests in central Illinois did not remain coniferous but accumulated other, less cold-tolerant species. Mixed coniferous and deciduous forests, likely with some pines such as what you find today in Michigan or Vermont, would have slowly spread northward. Eventually, coniferous species would have become less abundant and have been replaced by oaks, hickories, and maples. Similar vegetational changes happened everywhere in the Northern Hemisphere during these massive rearrangements of climate and earth.

Glaciers were not the only climatic driver of vegetation dynamics that this continent experienced. There was also a prolonged period of warmer conditions between 5000 and 9000 years ago that resulted in the expansion of the central grasslands. The Prairie Peninsula, postulated by Edgar Transeau in 1935, developed during this time as a broad swath of grasslands that pushed eastward from the central grasslands into Indiana and Ohio. The warmer conditions that dominated this period made fires more frequent and severe, tilting the ecological balance from a tree dominated landscape toward one dominated by grasses. With their adaptations to being grazed grasses keep their growth points low, whereas trees and shrubs can lose all their aboveground growth points in a severe fire. When there is a local fire, woody plants bear the brunt of the severe damage, but grasses typically grow back more abundantly following fires. If fires are frequent enough, trees are lost from the landscape, restricting them to wetter areas along rivers where they are protected. If fires occur often enough to reduce but not eliminate tree regeneration, you get an open-canopied savanna—the transition zone between prairies and forests. Infrequent fires are an essential component of many contemporary forests; it is harmful to remove fire from systems adapted to it.

The record of past fires is often set into the trees themselves. Every time there is a low-intensity surface fire in a forest, some trees will develop a fire scar. These scars are often on the tree's uphill side, where leaf litter accumulates and allows the fire to burn long enough to damage the bark and some underlying wood. With a careful accounting of the growth rings, scars re-create a location's fire history. Native peoples often burned the forest's deep leaf litter layers to encourage the tree species they liked, keep the forest more open, and drive game animals. Such fires undoubtedly affected the landscapes that we often consider to be the *original* forests of Eastern North America. The lack of fires influences forests today. The continued effects of people, both indigenous and European, on Eastern forests make a discussion of what truly is natural a difficult task.

The combined forces of environmental change, weather fluctuations, and human actions make species ranges much more dynamic than is often appreciated. Once we make a map, we tend to view it as immutable. All the large-scale changes discussed above influenced trees and other species at the time, and similar processes are still at work. Even mammals are subject to this problem. Two species of tropical origin, the nine-banded armadillo (*Dasypus novemcinctus*) and the Virginia opossum (*Didelphis virginiana*), have been moving northward aided by human alteration of the landscape and, more recently, climate change. If your field guide is more than a few years old, it is no longer valid! Although much slower and often less noticed, forest trees also move generationally. Removal of fire from grazing lands and prairies has resulted in a marked expansion of junipers (*Juniperus* spp.) from areas naturally protected from fire into fire-prone, grass-dominated areas. Usage of Osage orange (*Maclura pomifera*) and black locust (*Robinia pseudoacacia*) as fence posts has made very narrowly distributed species become common across broad swaths of the continent. These are all native tree species; species introductions and invasions represent new tree species still sorting out their ecological limitations on our continent.

## The Rate of Change Is the Difference

The previous sections have shown that forests change—species go extinct, continents bounce around on the planet's surface, and the climate changes. There is no denying it: forests are inherently dynamic. These factors have shaped the forests and other plant communities in North America today. The forests have changed continually over their long histories and will continue to change in the future, even without further human influences. The question then remains, why should we care about the changes happening today when change in forests is common? Quite simply, it is a question of temporal scale.

The evolution and assembly of the continents and their floras took place over millions of years. The continents did not shift in one or a few tree generations but in thousands upon thousands of generations of trees—more than enough time to see the evolution of new tree and pathogen species. Similarly, the last glacial maximum was 20,000–25,000 years ago, with the retreat and vegetation development occurring for thousands of years as the climate changed. Even the dry period that resulted in the expansion of the central grasslands ended 5000 years ago—that's twenty-five tree lifetimes if we assume that a tree lives on average 200 years.

This book focuses on forest changes, on losses of species, that occur much more quickly and, to date, are more irreversible than any natural threats forests have faced. The five trees that constitute the core of this book are facing, or have faced, functional elimination from forests within a single generation of trees—a time course of decades, not centuries or millennia. Global climate change, not the topic of this book but still a significant issue, will have much slower and persistent influences on plants and animals. The rate of climate change appears to outpace the ability of long-lived organisms like trees to move or evolve in response. The challenges to the forests of Eastern North America, and large portions of the planet, are happening on such brief time scales that they overwhelm the abilities of trees to evolve.

## The Way Forward

The following chapters will illustrate the changes in our forests over the last century or so. Within these chapters is described the unrelenting damage that our forests have suffered. Most of these stories include species invasions from elsewhere that disrupt populations that have not evolved the ability to deal with the pests. The loss of forest species comes with associated economic and ecological costs, many of which remain unknown or are perhaps unknowable. While these losses are serious and disruptive, they also provide opportunities to see nature in action. Forests are dynamic in space and time and will continue to be so. Forest composition will change, but the forests will remain as forests despite the losses of component species. Altered forests will hopefully still provide many, but perhaps not all, of the same ecological and socioeconomic functions.

Evolution will continue to happen within affected tree populations. We have introduced massive selection pressures on some tree species through our actions, resulting in the loss of many susceptible species. Strong selection pressures, such as those imposed by pesticides or non-native pests, can rapidly lead to evolutionary changes. These forces have recently generated the repeated evolution of herbicide-resistant weeds across many farms. If the only weeds to survive herbicide application are the few with herbicide resistance, the next generation of weeds will be dominated by individuals with genes for herbicide resistance—evolution in action! Though slower to respond to selection because of their longer life spans, trees face similar selection pressures and may also evolve resistance to their threats. In many cases, dedicated people stand ready to assist the evolutionary process in any way they can.

ULMUS AMERICANA. L

Leaves, flowers, and fruit of American elm, *Ulmus americana*.

# 1

# AMERICAN ELM — *ULMUS AMERICANA*

Trees provide some of our most common street names—after numbers, presidents, and words like main or high. Chief among these tree names is elm, though the likelihood that you have ever walked along a street lined with elms is slim. American elms made excellent street trees because of their elegant form: they grew tall and straight naturally, with branches that angled sharply upward. With little pruning, they would produce graceful arching canopies over the sidewalks, streets, and streetlights that dominated American towns. Many towns relied on American elm as their primary street tree because of this growth form. Though American elms are mostly long gone, they persist in the street names they left behind.

Because of its massive size and graceful shape, the American elm has always been natural gathering places for people. Under its shade, countless picnics, family reunions, political discussions, and religious gatherings have occurred over the centuries. Some of these trees have received names, solidifying their role as a local historical landmark. Many societally important elms have been lost over the years, leaving large holes in the communities where they once occurred. Examples of famous American elms include the Treaty Elm of Philadelphia (died in 1810 in a storm), the Liberty Tree of Boston (cut down in 1775 by a British loyalist), and the Logan Elm of Circleville, Ohio (weakened by Dutch elm disease, it died in a storm in 1964). The impressive size and scale of a large American elm are impossible to convey if you have never seen one of the remaining giants. The elm forests that greeted the first settlers must have been amazing to behold—some trees 5 feet (1.5 m) in diameter with trunks clear of branches 40 feet (12 m) or more from the ground.

I can remember the first time that I saw an American elm. I was in a dendrology class, and we walked up to a large, sickly-looking tree. It sported the stumps of several large branches that had been removed, and a pallid white streak that went down the main stem. We took notes, touched the leaves that separated it from its much fuzzier and less impressive member of the same genus, slippery elm (*Ulmus rubra*), and walked away completely unimpressed. More than two decades later, I saw what I would consider a "real" elm. My wife and I were walking with friends in Sacramento, California, when I stopped to notice the trees along the street. They were tall, nearly 3 feet (1 m) in diameter, and arched elegantly over the road. They also had the spongy, cushiony bark that characterizes elms. I had to examine a few trees before I found one with branches low enough to touch a leaf and verify they were American elms. I also found a series of holes drilled into the trunks, showing that someone had recently treated them for one of the diseases that has wiped them out of the forests of Eastern North America.

What happened to the American elm? It met ecological challenges that it could not rise to—new enemies in the form of diseases—Dutch elm disease and the elm yellows. Evolution is a beautiful thing, but sometimes it is just not fast enough to accommodate rapid changes to an organism's habitat. Could the American elm have adapted? Absolutely, and there is evidence that some trees may have become resistant to the diseases. The more significant issue is whether a once-common tree can survive a cataclysmic reduction in its population and regain its dominance across the range of habitats it once occupied. The unlikely probability of that happening is similar to that of a new species spreading across the landscape and becoming a dominant part of the forest community (chapter 7). Yet, many invasive species have achieved dominance in North America, so unlikely does not mean impossible.

Elms are not the first tree that we lost from Eastern forests, but they should have been the most visible in our day-to-day lives because of their central role in shading many of our towns and cities. Despite that visibility, the American elm is a tree whose loss seems to have gone unnoticed by most people, other than those needing to replant a great many street trees. We seem to have missed this lesson, however. Diversity builds resilience, even among neighborhood trees. The current threat to ash trees (*Fraxinus*; chapter 4), like the widespread loss of the American elm from the 1950s, might also require the large-scale replanting of street trees. The real ecological cost of losing the American elm can only be guessed, as its loss predates the ecosystem-level approach necessary to gauge the effect adequately. However, what is clear is that the loss of this species will affect us and our forests for generations to come.

## The Ecological Role of Elm

As with all elms, American elm can grow quickly and spread rapidly when given the right conditions. They are wind pollinated, flowering and releasing their pollen before their leaves have expanded in the spring. Most wind-pollinated trees bloom early in the growing season when there are fewer obstacles to pollen movement. Releasing pollen into the wind is a very messy and inefficient way to father seeds. There is no way to direct the pollen to another individual of the same tree species, let alone toward a flower. To make up for this, wind-pollinated plants produce copious amounts of pollen to increase the chances of successfully pollinating a flower. Most of this pollen will still not travel great distances, decreasing in abundance exponentially as it moves farther from the plant that produces it. So, while the pollen can go great distances, the most reliable pollination will occur over close distances, on the scale of 100 feet (30 m) or so. The most significant benefit to wind pollination is the reliability of wind, and the plant doesn't need to spend any energy to encourage the wind to move pollen.

Elms produce seeds a few weeks following pollination that, like the pollen, are also spread by the wind. The tree's seeds are relatively small, about the diameter of a lentil but much thinner. The fruit that carries the seed is a papery disc of tissue that acts as a wing to keep the seed afloat in the wind. Botanists refer to this fruit type as a samara. The samaras are released in the spring and can move appreciable distances, even more so if they land in water so that spring floods can move them downstream.

Because of their small seed size, elms commonly colonize openings in the forest canopy or open areas adjacent to forests. Small seeds cannot contain sufficient energy reserves to grow and persist in shaded conditions. However, once established, elm trees have the capacity for fast growth and can live for a long time. This combination of characteristics—good dispersal, small seeds, and rapid growth—makes American elm a colonizer of open, disturbed habitats. The ecological gap left by the loss of such trees is too often filled by non-native trees with weedy life histories. We will discuss threats to another tree that commonly colonizes disturbed environments, the ash, later in this book (chapter 4).

Our best information about the historical composition of forests in Eastern North America comes from a foundational book by E. Lucy Braun (1889–1971), a plant ecologist based at the University of Cincinnati. Her *Deciduous Forests of Eastern North America*, based on decades of fieldwork for which she and her sister visited and described mature forests

Historical distribution of *Ulmus americana*. From U.S. Geological Survey Geosciences and Environmental Change Science Center.

across the eastern states, was published in 1950. They risked run-ins with moonshiners and others who would have shot men walking through their forests without question but tolerated two women who just wanted to look at forests and knew how to keep their mouths shut. Braun described many of the remaining old-growth forests before they were cut down and lost forever, making her work an invaluable and lasting contribution to science. She organized this information into a comprehensive guide to the many forest types that occupied Eastern North America's diverse land-scape. Her work represents our best, and sometimes only, view of what forests looked like before we cut them down, removed their natural fire regime, and released plagues of insect pests and diseases upon them.

American elm appeared as a relatively minor component in many of the forest types that Braun described, often constituting only 1–3% of the canopy trees in upland forests. In contrast, American elm became much more abundant in lowland forests with moister soils. In swampy forests or floodplains along rivers, American elm could make up one-third or more of the forest. This dominance was prevalent in areas where disturbances had removed other trees, allowing elms to colonize and quickly dominate the area. Braun considered the American elm to be primarily a succes-sional species, which would ultimately be replaced by more shade-tolerant species such as sugar maple (chapter 5). However, the species' interme-diate shade tolerance also allowed it to grow in closed-canopy forests, likely dependent on large treefall gaps that provided the necessary light to become established. American elms also live much longer than many early successional trees, so they do not fully represent the live fast, die young strategy of the earliest successional species.

American elm's topographic distribution may be partly responsible for the lack of public notice or awareness when the species began to disap-pear from forests. In uplands, the species was only a small part of the tree canopy, so the loss of a few individuals would not represent a dramatic forest change. To place this damage in an appropriate ecological context, forests typically have 1–5% of their canopy in gaps at any point in time. When elms die, the canopy space they leave behind may be taken over by adjacent canopy trees' expansion. Larger forest gaps may allow new tree seedlings to become established and eventually fill in the canopy. Inter-estingly, when large American elms die, they generate the sort of open conditions for which they are well suited. One study found that the loss of large canopy elm trees resulted in a flush of new elm seedlings becoming established, increasing the total number of American elm trees over time. Of course, the newly recruited trees were much smaller than those which formerly occupied the canopy. Furthermore, they would be unlikely to survive long enough to reproduce more than a few times. This pulse of recruitment, while interesting, is likely just a transient feature of the forest and should still ultimately lead to the replacement of American elm by other species. In other habitats, the loss of canopy elms resulted in an explosion of shrubs in the now illuminated forest understory. Competition from shrubs may inhibit tree regeneration for decades, resulting in a forest with persistent openings until the shrubs ultimately die.

In wetter lowlands where the American elms dominated, the forest canopy effects would have been much more significant. Large numbers of dead elms would have allowed the rapid expansion of other early successional species that also do well in wet conditions. Species expected to do well under those conditions include willows (*Salix* spp.), cottonwoods (*Populus* spp.), boxelder (*Acer negundo*), and silver maple (*Acer saccharinum*). As these species grow even more quickly than American elm, forest canopy recovery would have occurred swiftly. As these colonizing species also have low shade tolerance, their persistence in the landscape should only be transient. Riverside habitats are also typically altered to improve drainage or otherwise make them more useful to us, so there may have been few forests that retained American elm dominance when Dutch elm disease spread in the middle of the twentieth century.

One place where the loss of elm has entered into our awareness is in the folklore of mushroom hunting, particularly the much-desired morel. Many morel hunters will seek out places where there are dead or dying elms (any species) as those habitats are supposed to produce more morels. Though I have heard several mushroomers describe the relationship between elm and morels, I have never heard any of them speculate whether the elms' dying is either good or bad, nor a discussion as to what might happen if all the elm trees die.

## Dutch Elm Disease

An unlikely and somewhat accidental partnership between fungi and insects drove the loss of American elm. Dutch elm disease is named so because it was described first in the Netherlands in 1921—please don't blame the Dutch. Confusingly, there is also a hybrid elm known as the Dutch elm (*Ulmus* × *Hollandica*). This elm is susceptible to the disease but got its name from the Dutch who initially developed and propagated the tree variety. It was not the source of the disease. The fungi that cause Dutch elm disease come from a suite of closely related species in the genus *Ophiostoma*, which are ascomycetes. Ascomycetes are named for the reproductive structures of the fungi. When these fungi sexually reproduce, they produce their spores in small sacs, known as asci. The disease-causing fungus's genus was formerly *Ceratocystis*, so if you look into the older literature, that name will appear. Such name changes are common in the shifting sands of taxonomy and hopefully reflect a better understanding of the organism. The ultimate origin of the fungal pathogen is unclear, although Asia seems most likely.

This fungus has different mating types, not fully male and female, as there are no morphological differences or variation in how they provision for their offspring. These mating types simply prevent them from mating with themselves. In Europe, populations tend to be much more diverse and contain multiple mating types, so finding a sexual partner is more common there. The genetic diversity that sexual reproduction generates maintains Europe's diverse pathogen population. In North America, there is much less genetic diversity because of the limited introduction of the fungus. On top of that, colonization of new patches of elms is a random event. Only one genetic individual of the fungus colonizes most American elm stands, so sexual reproduction rarely occurs within the invaded range. This lack of sexual partners for the fungus means that spore production through asci, and release into the environment, also does not occur. So how does the infection spread? A vector, something that moves the fungus from tree to tree, is needed.

Let's take a moment to discuss woody stem anatomy. The innermost portion of a tree trunk is the wood (xylem) that supports the tree and conducts water. The main water-conducting cells are dead at maturity but are attended by living cells to maintain their function. In larger stems, the centermost wood will no longer be functional for water transport as the attending cells in that area will be dead. The outermost region of wood, often called sapwood, is actively involved in water transport as it retains living cells. The sapwood is surrounded by the cambium, the growth layer for the tree stem. Cambium tissue produces wood to the inside and phloem, sugar-conducting cells, to the outside. Just to the outside of the cambium layer will be the inner bark, which will also be alive at maturity. So, in a large tree, the only stem parts that contain living cells will be several years' worth of sapwood, the cambium (a single cell layer), the phloem (maybe dozens of cells), and inner bark (dozens of cells). This relatively narrow band also represents most of the nutritious tissue in a tree stem.

When a stem becomes infected with one of the *Ophiostoma* fungi, the fungus grows through that narrow layer of living tissues, digesting its way through the stem. In response, the elm begins to block sapwood cells with gums and cellulose. While this vain effort does little to stop the fungal spread, it rapidly reduces the plant's ability to move water from the roots to the top of the tree. Blockage of vascular tissues rapidly leads to wilting and yellowing of the leaves in the canopy, followed by crown dieback and eventual tree death. In cases in which there is a heavy infection, tree death can occur in as little as a year.

But how does infection occur in the first place? Without assistance, a nonmotile spore would need to get past the protective outer layer of bark and make contact with the living inner tissues. Wounding of the stem could allow this, but there is an active player, a bark beetle, that lends assistance. Bark beetles are a diverse group of small insects that forage on trees' cambium, phloem, and inner bark. Bark beetles don't penetrate much into the sapwood as food availability there is much reduced. Therefore, they spend their lives in that same thin plant layer used by the fungus. This stem location is an excellent ecological niche as it is a good food source and the bark protects the beetles, making it hard for predators such as birds to find their prey. In western North America, many conifer species are being affected at plague levels of both native and introduced bark beetles. Wherever bark-foraging beetles become abundant, it is never to the benefit of their host trees.

There are three elm bark beetle species, so named because they are primarily elm specialists, though they can occur on other tree species. These are the native elm bark beetle (*Hylurgopinus rufipes*) and two European species from the same genus, the European elm bark beetle (*Scolytus multistriatus*) and the banded elm bark beetle (*S. schevyrewi*). *Scolytus multistriatus* has been present in North America since at least 1910, well before the first incidence of Dutch elm disease, so that beetle's abundance was more of a predisposing factor in the loss of American elms than a direct cause. All of these beetles prefer to lay their eggs under the bark of elm trees, particularly in those individuals that are stressed from drought or disease as they may have fewer remaining defenses. The female beetle burrows into the bark and chews out a long tunnel, laying her eggs along the sides as she goes. When the eggs hatch, the larvae chew out little tunnels running out from the maternal tunnel, producing what looks like an exaggerated cave painting of a centipede with exceptionally long legs. Eventually, the larvae mature, chew their way out, and find another elm to host the next generation's eggs.

Since the bark beetles preferentially select stressed trees, they often choose trees to lay their eggs in that are experiencing Dutch elm disease. When the fungus is present, it fills the beetle's tunnels with fungal hyphae and asexually produced spores, effectively cloning itself. When mature, adult beetles emerge from under the bark covered with sticky spores. As the young beetles mate and burrow into new elms to lay eggs, they infect their host trees, further weakening the tree and increasing the beetle's success. This interaction between beetle and fungus results in positive feedback between the two players. Stressed elms attract bark beetles, produce lots of new beetles that spread the fungus to new hosts, weakening more trees, and so on until most of the elm population rapidly becomes infected.

The rapid spread of the disease and selectivity of the beetles explain some of the odd tree population dynamics that result from the disease. Mature trees with bark thick enough to provide sufficient food for developing larvae and protection from winter cold are attacked and may quickly die early in an infestation. Smaller trees with thinner bark are not good hosts, and so the population of bark beetles may decline following the initial pulse of elm mortality. As the elms die, so does much of the fungus, and the potential for infection declines. The remaining small elms in the forest may grow for years before they get large enough to produce bark suitably thick to support bark beetle reproduction. By that time, there may be no beetles in the area, or those around may not carry the fungus. Sooner or later, however, the fungus and beetles reconnect and generate a second outbreak within the forest. If the American elms can grow large enough to reproduce for a few years before the disease catches up with them, another generation of young American elms may be produced. This cyclic process may repeat itself time after time within a forest patch. Many forests retain smaller American elms as lesser components of the canopy and subcanopy—persistent shadows of their former glory.

## Dutch Elm Disease Progression in North America

A few fungal species are responsible for Dutch elm disease, and the individual species have played different roles at different times in the disease as it progressed. Dutch elm disease was widespread in Europe when it appeared in Ohio in 1927, apparently imported on elm logs for veneer in the furniture industry. I'm not entirely sure why elm wood would need to be imported, given the number of native trees available. Elms tend to form burls, large tumor-like growths with a swirling grain pattern prized in furniture making. Perhaps imported burl was considered more attractive or otherwise preferable to domestic burls. Large burls would also provide lots of nooks for beetles to go unnoticed and are valuable enough that no one would turn them away because of a little fungal damage. Of course, there may have also been a surplus of burls available from salvage logging in Europe as their trees died from Dutch elm disease, providing a ready source and incentive for export. The great irony is that had the logs been converted to veneer before importation, the likelihood of disease transmission would have been much, much less. The fungus would have still been present but had little chance of meeting up with its bark beetle vector.

Regardless of the economic drivers that resulted in the wood movement, the logs were imported and carried with them the first of the Dutch elm disease fungi (*Ophiostoma ulmi*). This fungus spread from its original point of entry, moved from town to town on logs from dead and dying trees that were cut down and transported. From infected logs, bark beetles flew out to find new trees to colonize. The way that we planted elms, in long lines along streets, also facilitated tree to tree transfer. Spread occurred through either new infections or the sharing of an existing infection through root connections between neighboring trees. While the spread and severity of the disease slowed down in Europe as the fungus picked up some viruses, the fungus's natural enemies did not occur in North America. The disease progressed in North America unabated, continuing to spread and kill trees. Though this first wave of infection was deadly to many American elms and certainly should be considered catastrophic, many American elms survived.

A second invasive fungal pathogen, *Ophiostoma novo-ulmi*, which translates to the new elm-infecting *Ophiostoma*, appeared on the scene in the 1940s. With different environmental tolerances, this fungal species was much more virulent than the first, killing the vast majority of the American elms left from the original disease outbreak. Interestingly, this second pathogen has two different strains: one from North America and one from eastern Europe, now thought to be a subspecies. In North

America, *O. novo-ulmi* started in the Great Lakes region and spread outward from there, by the 1980s covering the continent except for some northern refugia. The spread of *O. novo-ulmi* effectively replaced *O. ulmi* in the landscape, though the original disease persists at low abundance.

Those who ignore history are doomed to repeat it. As we ignored the original ecological lessons about transmission pathways from Dutch elm disease, a second infection has occurred. The North American strain of *O. novo-ulmi* made it back across the ocean with exported elm logs and has spread throughout Western Europe, killing trees.

## Insult to Injury—Elm Yellows

One catastrophic disease would seem to be sufficient to decimate American elms, or at least to set them back dramatically in the modern landscape. However, there is a second pathogen that affects this and other elm species—elm yellows. As plant disease names are descriptive, this disease characteristically results in the yellowing of elm foliage and causes rapid mortality even in large trees. There is no cure, so management focuses on preventing infection. This is a polite way of saying that trees showing symptoms and those surrounding them must be immediately cut down and disposed of sanitarily. Problematically, elms that are being treated for Dutch elm disease rapidly succumb to this second disease.

The disease is caused by a phytoplasma, a bacteria that is an obligate parasite on plants. The elm yellows pathogen infects the phloem, the sugar (i.e., photosynthetic energy) conducting tissue of a stem. Once the phytoplasma gets into the phloem, there are plenty of resources for the bacteria to grow and spread, cutting off the sugar transport throughout the stem. The disease functions much in the same way that Dutch elm disease does, though the phytoplasma are restricted to the phloem. The tricky thing for the bacteria is getting into the phloem in the first place. Just as Dutch elm disease has an insect vector to move it around, so too do the bacteria causing elm yellows.

Because of the high caloric and nutritional value of the materials carried in the phloem, many specialized insects feed on this layer. Some insects physically consume the phloem, as does a bark beetle, but many take a more pragmatic approach. Plants actively load sugars and other compounds into the phloem to get them quickly distributed from where they are produced, typically leaves, to where they are used. So the phloem, like a human's vascular system, is pressurized to enhance flow. All an insect needs to do is insert its mouthparts into the phloem, and the plant will move resources to it—no chewing, moving, or other insectly work required—the equivalent of a tick on a mammal. The most familiar of phloem-feeding insects are the aphids, small, soft-bodied insects that can reproduce asexually. Phloem feeding generates little to no physical damage to the plant—perhaps only a pinprick brown spot where the insect fed. The sinister thing about phloem-feeding insects is that they can pull off so much of the sugar produced through photosynthesis that they dramatically reduce plant growth.

Phloem-feeding insects are famous as vectors for plant diseases, just as mosquitos and ticks are for blood-borne diseases in humans and other mammals. Insects move from plant to plant, picking up a bacteria or virus from one plant, obliviously injecting that pathogen into the next plant upon which they feed. As many phloem-feeding insects are specialized feeders on one or a few plant species, they also tend to move disease from susceptible plant to susceptible plant. The chief vector of elm yellows is the North American native white-banded elm leafhopper (*Scaphoideus luteolus*). Leafhoppers can be quite colorful but many hide in plain sight, camouflaged like a plant thorn that may hop away if you approach them. These insects are so common that disease transmission is best reduced by preventing insect vectors from picking up the pathogen from infected plants rather than by directly controlling the insect.

## The Future of American Elm

Despite the destructive consequences of the European and American diseases on the American elm, some American elms still persist, particularly as populations become more isolated in the Northern Plains. Several Canadian cities maintain American elms as street trees, but they need to be constantly vigilant against the first case of Dutch elm disease or risk losing them all. Arboreta, college campuses, and some cities, such as Sacramento, maintain trees in this way.

From a street tree perspective, there are some replacements for elms. Most similar to the American elm's characteristic shape is a tree called

Japanese zelkova or keyaki (*Zelkova serrata*). This species has the right shape, but as a non-native tree the last thing we need is another forest invader. Similarly, Siberian elms (*Ulmus pumila*) have a much weedier life history than American elm and are used for windbreaks and similar uses in the upper Midwest. This invasive species has become established in many states but does not fill the same niche as the majestic tree that we have lost. Some American elm varieties have resistance to the fungus and show preliminary promise for reestablishing street tree populations. The earliest and perhaps most prominent of these are the Princeton elms, named after a famous line of elms trees that have persisted blight-free for decades in Princeton, New Jersey. Propagated by cuttings, these trees are thus clones of the originals and are available in specialty nurseries. I have three of these in my yard on my farm in Illinois. As of this writing, they are 4 years old, over 16 feet (5 m) tall, and want to grow upright more than any tree I have ever seen. They are simply beautiful.

There are other American elm varieties and hybrids that are reputed to be Dutch elm disease resistant. They have inspiring American names like "Valley Forge" and "Patriot." As modern horticulture prizes plants that can be copyrighted, patented, and monetized. I would expect to see others developed, particularly if there is interest. I've seen disease-resistant elm varieties planted in Cincinnati, but they were young and newly planted. Whether the street trees of yesteryear will rise to the chemical and physical challenges of modern cities is unclear, but it is great to see the effort attempted. Problematically, many American elm varieties are all still susceptible to the elm yellows. Some hybrids are allegedly resistant to both diseases, but moving toward hybrids shifts us even further away from the regal and genetic stature of our native American elm. It is difficult to know how hybrid trees would affect native forests if they were to become naturalized in the landscape.

Several groups are working on breeding disease resistance into new varieties and have established restoration plantings along floodplains, where American elm would have been dominant. The hope is that by generating a variety of disease-resistant trees, they can establish new populations that will produce disease-resistant offspring, able to persist without further intervention. The genetics of disease resistance is difficult, as any resistance to the disease also represents selection pressure on the disease to escape that newly acquired resistance. As the diseases have much faster life cycles than the trees, the evolutionary advantage distinctly lies with the fungi and bacteria. Disease-resistant populations will need to have enough variation in their resistance mechanism to allow for continued selection to operate and counter any evolutionary changes in the pathogens over time. Simple, one-gene disease resistance can be quite effective in the short term, but suppose the disease organism has some genetic variation that allows at least some to infect resistant plants. In this case, there can be a rapid and catastrophic selection of a new disease line that bypasses plant resistance. Selection for a new disease genotype would effectively reverse all progress initially made. However, the sparseness of elms in the landscape may dramatically slow the spread of any resistant fungal genotype. Ideally, multiple resistance genes with different mechanisms of action will be developed so that the fungus would need to have more than one virulence gene to overcome plant defenses.

When it was widespread, American elm undoubtedly was a genetically diverse species. While that diversity was insufficient to save it from Dutch elm disease and the elm yellows, genetic diversity will likely be important down the road. The long-term fear is that elm reestablishment will rely on a few genetic individuals or small pools of closely related individuals that share disease resistance—something referred to as a genetic bottleneck. Genetic changes like this mean that the established populations will also tend to have low diversity. Bottlenecks are one of the main concerns in protecting any endangered species. It is possible to save the species only to watch it succumb to internal genetic challenges posed by low-diversity populations. Problems can arise from limited potential to adapt to new environments, climate change, herbivores, or pathogens. Challenges can also come from inbreeding and the accumulation of harmful gene combinations. This is the reason you don't date your close cousins, or at least have children with them. Concerns and worries are not a reason not to try breeding resistant elm varieties, but rather these concerns assist in identifying problems to be vigilant against as we proceed. Species may pass through genetic bottlenecks and regain much of their diversity, but this does not always happen.

Apart from genetic diversity, the primary challenge to effectively restore a species that once occurred across the eastern half of North America is spatial scale. Once resistant lines are developed and effectively breeding, they will need to be established in hundreds if not thousands of sites. Seed dispersal in American elm is quite good, but not so good as to colonize half a continent. By planting populations along rivers, it may be possible to establish source populations that could then disperse long distances via the waterways. As dispersal would mostly be downstream and less likely as distance increases, initial plantings would need to be spaced according to reasonable dispersal distances in rivers. Established individuals within local habitats could expand populations as nearby trees die and potentially contribute to further downstream dispersal. Of course, rivers only flow one way, so each drainage's headwaters would need to have populations installed to colonize all former habitats. Luckily, American elms reproduce rapidly and early in their lifespan, which would help to reduce the time necessary for reestablishing populations.

As a planning exercise, reestablishing a species over such a large area seems at first too daunting to even start. However, we have strong precedents for this approach, although in a much less planned way. The successful invasion of the myriads of non-native plant species (chapter 7) suggests this is not only possible but something that may be almost impossible to stop once it has begun in a successful species. Siberian elm is commonly planted as a windbreak tree and has become established in almost every state and province of North America. You typically find the species in abandoned lots in towns, hedgerows, and other places where trees can grow uncontrolled. If we have accidentally spread a non-native elm species across the whole continent in the space of several decades, what more could we achieve with a beloved native species with a little planning and forethought?

It is beautiful to think that restoration for the American elm may be as simple as replanting them as city trees. Cities and towns are spread throughout the landscape, in contact with natural areas, rivers, and almost every habitat imaginable. Planting and taking care of these trees will generate a massive reservoir of seeds that land in hedges, garden beds, and the back corner of the lot that you haven't quite gotten to yet. Year after year, a few trees may be able to grow and reproduce on their own. Some seeds will eventually make it out of town, stuck in your car's windshield, washing down the town stream, or carted off with lawn waste. A minute number of seeds may become established, grow, release seeds, and start new populations. We have watched this process with hundreds of non-native plant species. A plant's capacity, particularly an early successional one, to reproduce and spread is fantastic and may work in our favor for once!

CASTANEA DENTATA, Borkh.

Leaves, fruit, and seeds of American chestnut, *Castanea dentata*.

# 2

# American Chestnut—
## *Castanea dentata*

During the sweltering California summer of 1945, Mel "The Velvet Fog" Tormé and Bob Wells wrote the audaciously titled "The Christmas Song." Not *A* Christmas song, but, in fact, *The* Christmas song, in opposition to all others. Perhaps such hubris would have faded into obscurity, but the Nat King Cole Trio recorded the song in 1946, resulting in the birth of a holiday favorite. For me, nothing marks the beginning of the holiday season quite like the first notes of Cole's rendition of this holiday classic song. I think it reminds me of my childhood, filled with the voices of Bing Crosby, Rosemary Clooney, and, of course, Nat King Cole.

Why mention a classic holiday song in a book on trees? It is the first line of that song—"Chestnuts roasting on an open fire." How many times have people heard or sung this line and never wondered what it refers to? As the actual title is so vague, the song is often subtitled with that first line. Perhaps Tormé and Wells were reminiscing on their childhoods in cooler climates with chilly winters when they penned the song. By the time they wrote "The Christmas Song," the American chestnut was already in serious trouble, and it is not likely that there were many chestnuts around for roasting. Our connection with the lyrics has diminished, accompanying our relationship to chestnuts, but the song is so ingrained in our collective conscience that the tradition remains.

Chestnuts make an appearance in another traditional holiday song, "Sleigh Ride." This piece was originally an instrumental composed by Leroy Anderson in 1948, with the chestnut-referring lyrics added in 1950 by Mitchell Parish. In this song, the chestnut reference is to gather around the fireplace "while we watch the chestnuts pop. Pop! Pop! Pop!" I cannot help but think that this contemporary of "The Christmas Song" is another recollection of the lyricist's childhood. When you roast chestnuts, you cut a slit in the outer shell to allow steam to escape. If you do not, they explode like popcorn, but an order of magnitude more violent. This pop also indicates when the chestnut is adequately cooked. Using such a timing device is quite dangerous based on internet warnings and is unnecessary in our modern kitchens with well-controlled ovens. However, in the unpredictable temperature of an open fire, leaving a few chestnuts whole creates a self-timer for cooking. When the chestnuts pop, the remaining nuts should be done—it is dangerously ingenious.

Obscure references in two songs are the limit of direct experience with the chestnut for most North Americans. Species related to our native American chestnut remain in Europe and Asia and are grown in North America. These species now serve as our source for chestnuts. While we no longer have a societal context for chestnut trees, they once occupied a significant portion of many forests, yielding food, timber, and structure for us and wildlife. Their loss set the stage for much of the forest structure that we have today.

## An Incredibly Useful Tree

One of the great tragedies of the death of a tree species is the loss of their multidimensional utility, for which the American chestnut is an excellent example. Chestnuts provided a critical food source for Indigenous Americans and the Europeans who displaced them. From a human history perspective, this is undoubtedly the chestnut's most significant contribution to our lives. People living in the southern Appalachians sold or bartered chestnuts for supplemental income, shipping train-car loads to consumers in the north. Chestnuts are primarily composed of carbohydrates and can be cooked and eaten fresh or dried and milled into gluten-free flour. Once milled, chestnut flour is used as any grass-based grain. Strong advocates for permaculture and woody agriculture have argued that the chestnut, particularly Asian species and hybrids, could easily replace corn as the primary carbohydrate crop, with accompanying environmental and ecological benefits. As someone who grew up in the corn belt, I have to wonder about farmers' thoughts on that transition!

From a materials perspective, chestnut timber is also incredibly valuable. The wood is strong, light for its strength, and rot resistant. These qualities made chestnut a choice timber for framing barns, homes, and industrial buildings. The large size of chestnut trees meant that sufficiently long and large beams could be produced for nearly any purpose before steel was available. The wood splits well, which in the past made it ideal for fence posts or shakes for roofing. The rot-resistant wood is also relatively stable, allowing it to be used effectively in building furniture, barrels, and pretty much anything you may want to make from wood. And, it should be noted, chestnut wood is beautiful. It has a strong grain similar to oak but with a natural warm brown color. The wood isn't chestnut brown, as that color refers to the reddish-brown of the nut rather than the wood. Instead, the wood naturally has the rich brown color that we often stain oak to produce.

Chestnut bark and wood contain large amounts of tannins, used in processing animal hides into usable leather. While we no longer use tree bark for this purpose, there is no denying the importance of this trade and its necessary base materials in human history. The production of useable amounts of tannins is not unique to chestnut, but the species was the primary source in the southern Appalachians. Traditional processing of hides in this way polluted streams horribly, so it is for the best that the industry has moved on to more environmentally friendly methods. Nonetheless, chestnut was an incredibly beneficial tree for a long, long time—and then it was gone.

## The Chestnut in Our Forests

Over the years, chestnut trees have moved from our shared common familiarity into the realm of legend and myth. Since the American chestnut was primarily a tree of the forest, not of cities, the chestnuts' abundance in forests has been artificially expanded over the years and now has captured the imagination of well-meaning forest conservationists and tree lovers. I have seen the chestnut referred to as the most dominant tree of the eastern United States, with suggestions that one in four trees in the Eastern forest was a chestnut, and similar comments meant to capture the enormity of the loss suffered by our forests. Others have railed against the exaggeration of chestnut abundance and point out how infrequent they were in many forested landscapes of New England—listing other tree species that are much more widespread and dominant across large areas. While chestnuts likely weren't as abundant as some would now assert, their loss is nevertheless genuinely tragic.

Chestnut's range had been most of the mountains and uplands of the eastern United States. The species thinned out in the northern mountains because of the colder climate and disappeared at the transition to the flatter, sandy coastal plain forests in the south. An inability to tolerate poor drainage kept chestnut out of lowland forests. It was in the forests of the southern Appalachians where the American chestnut achieved its greatest abundance. The Braun sisters (chapter 1) carried out their fieldwork when only the tall trunks remained of the recently dead trees. These trunks, however, along with data pulled from older surveys, allowed the Brauns to estimate chestnut abundance relative to other tree species. Along ridges in the Cumberland, Allegheny, and other mountains of the southern Appalachians, chestnut was often codominant with one or more oak (*Quercus*) species, forming chestnut–oak forests. Sometimes chestnut was the clear dominant in all senses of the word. The leaf litter produced by the chestnuts each fall did not decompose quickly, leading to a forest understory that tended to be much more open and supported fewer herbaceous species than other forests. American chestnuts had the potential to become massive trees. These forest goliaths regularly reached proportions of 4 to 5 feet (1.2–1.5 m) in diameter; the biggest individuals could reach 10 to 12 feet (3–3.7 m) or more in diameter. One of my wishes is to have walked among uncut chestnuts in the southern Appalachians. Old U.S. Forest Service photos show loggers dwarfed next to the trees they would soon cut and mule teams in front of logs stacked impossibly high, making you wonder how the loggers loaded them, let alone how the poor mules would move them. Most photos capture only a few of these standing trees because of their immense size. These forests must have truly been a wonder to behold.

The chestnut was also a component of the mixed mesophytic forest. Mesophytic refers to plant communities that are moist, neither too dry nor too wet, an environment suitable for a broad range of forest trees. Mixed is a much simpler term; these forests are incredibly diverse. You would need to travel southward until you encounter tropical forests to find a greater diversity of forest canopy trees. Relatively few species dominate most forests in temperate zones. We have oak–hickory, beech–maple, oak–pine, and spruce–fir forests, to name a few forest types from Eastern North America. Other species occur within these forests but at much lower densities relative to the clear dominants that give the forest types their names. The mixed mesophytic forest, in contrast, has no dominant

The distribution of American chestnut was centered along the Appalachian Mountains from northern Alabama, Mississippi, and Georgia to southern Maine. From U.S. Geological Survey Geosciences and Environmental Change Science Center.

species but rather a large suite of canopy trees with no species consistently achieving a numerical advantage over any other species. If you have ever visited the Great Smoky Mountains, you have seen the mixed mesophytic forest or at least a good portion of what remains. These forests have thirty or more canopy species, including maples (*Acer* spp.), oaks (*Quercus* spp.), ashes (*Fraxinus* spp.), yellow buckeye (*Aesculus octandra*), beech (*Fagus americana*), magnolias (*Magnolia* spp.), basswoods (*Tilia* spp.), tulip poplar (*Liriodendron tulipifera*), black cherry (*Prunus serotina*), and, at one time, chestnut (2–30% of stems according to Lucy Braun). As one moves to higher elevations, latitudes, or westward to drier sites, subsets of these species dominate.

Farther northward and westward from the mixed mesophytic forests, the chestnut became less abundant, persisting as a minor forest component. However, the chestnut could increase rapidly in the open sun conditions that followed logging. In these forests, chestnut made up a much smaller proportion of the forest, often 5–10%, though it could still be a locally dominant tree depending on land-use history. Again, these were massive trees, so the biomass they contributed to the forest was still relatively large. When the forests lost their chestnuts, they lost a sizeable proportion of their canopy, along with the photosynthesis, nutrient retention, soil stabilization, food resources, and shading the chestnut contributed. These functional losses likely persisted for decades, recovery occurring as surrounding trees grew and new trees slowly established from seeds to fill in the gaps left behind. Furthermore, the food resources represented by chestnut seeds, which many forest animals consumed, may have never recovered.

Is there a biological explanation for why American chestnut was so abundant across such a wide range of habitats and forests? In diverse forests such as North America's mixed mesophytic, tree diversity is often attributed to one species' inability to outcompete the other species because of fine-scale variation in the environment. To a tiny tree seedling, a forest understory is a very heterogeneous place. Conditions such as light, moisture, and soil fertility can vary dramatically from one location to the next. Trees are more sensitive to environmental conditions, herbivores, and pathogens at the seedling stage, the most critical stage in a tree's life history. Greater tree diversity occurs when one species is better able to regenerate under a particular set of environmental conditions. In contrast, another species would specialize on a different set of requirements. Variation in tree seedlings' performance across the forest floor allows the coexistence of multiple species—each regenerating somewhere slightly, but critically, different. If one tree species regenerates well across a wide range of conditions, it should increase dominance by that species and represent an overall loss of tree diversity.

In what I consider to be one of my favorite wish-I-had-thought-of-that studies, ecologist Roger Latham grew a suite of tree seedlings under various environmental conditions to test the importance of understory environmental variation in controlling tree diversity. In a greenhouse, he grew seedlings of hickory (*Carya tomentosa*), red oak (*Quercus rubra*), beech (*Fagus grandifolia*), black gum (*Nyssa sylvatica*), tulip poplar, and chestnut. He grew seedlings in three levels of light and nutrients in all nine possible combinations—with high, medium, and low levels each of nitrogen, phosphorus, and potassium—and measured a host of plant growth characteristics. As these tree species all coexisted in mixed mesophytic and other forests, we would expect that there should be some combination of environmental conditions that make each species superior, at least somewhere. What he found was the exact opposite. Chestnut performed better overall across a wide range of conditions, making the species less specialized in where it regenerates and more competitive in nearly all environments. This experiment presents a simple physiological explanation for the abundance and potential dominance of chestnut but doesn't explain how so many species can coexist. If anything, the experimental results suggest that chestnut should have been even more dominant across the region. There are, of course, a diversity of hypotheses postulated to explain tree species diversity in forests, and environmental segregation is just one. Beyond physiology, there are plant–soil microbe interactions, pathogens, herbivores, droughts, and winter damage—all of which may be important for tree regeneration.

Two physiological aspects of the chestnut that Latham's experiment did not address are soil moisture and pH. Moisture-wise, chestnut is also a mesophytic species. Chestnuts have relatively high water demands—something a greenhouse study cannot easily address. Chestnut's need for large amounts of water translates to a historical lack of chestnut along dry ridgetops and other exposed sites where rainfall would drain away too quickly and generate water stress in most years. At the other end of the moisture spectrum, chestnuts cannot survive without good drainage, so any places where water accumulated for periods chestnuts did not grow. The optimal moisture range for chestnut was between these two environmental extremes.

Soil conditions may be one reason why chestnut returned more slowly than many other forest trees of Eastern North America following the receding glaciers. Documenting forest recovery after glaciation utilizes a unique, almost archaeological line of ecological research. Pollen, particularly that of wind-pollinated species, is produced in large amounts. In areas where pollen can accumulate undisturbed, such as ponds or bogs, you get a continuous record of the species changes that occur over time, the oldest forest communities represented on the bottom, the more recent ones toward the top. If this approach is combined with a way to date the pollen layers, a local forests' compositional history can be reconstructed. The pollen of American chestnut appears in northern deposits in low amounts relative to other tree species such as birches (*Betula* spp.), beech (*Fagus americana*), and oaks approximately 4000–2000 years ago but then goes through a rapid expansion. As the species' pollen is found throughout the sediment strata, the chestnut dispersed northward along with the similarly large-seeded oaks and other tree species. However, it appears that something in the environment was constraining the species' abundance relative to other tree species following the glacier's retreat. Regardless of the impediment, it dissipated, and the American chestnut's competitive superiority allowed it to expand rapidly throughout its historical range and was reflected in pollen deposition. Much later, the decimation of chestnut populations by blight was also reflected in the pollen record. The regenerative capacity of chestnuts, above anything else, gives me hope for the species in the forests of our future.

### Chestnut Reproduction—Incredibly Edible Seeds

The regular production of large numbers of good-sized seeds represents the primary ecological role chestnut played, affecting both human and wildlife populations. Outside human consumption, chestnuts were important wildlife food sources for a wide variety of species. Most affected because of their inability to move great distances are the small mammals—chipmunks, mice, and squirrels. For large parts of the year, small mammal diets are composed primarily of seeds. Diets now center on oak acorns, but chestnuts would have originally been a significant source in many forests. Larger vertebrates such as white-tailed deer (*Odocoileus virginianus*) and turkey (*Meleagris gallopavo*) also fed extensively on chestnuts when available. The most significant difference between oaks and American chestnuts is the regularity of seed production. In marked contrast to chestnut, the oaks exhibit considerable year to year fluctuation in acorn production. Therefore, oaks represent a dramatically changing resource base for the forest food web with high production years, known as mast years, followed by one or more years of minimal reproduction.

Understanding pollination in chestnuts is tricky as there does not seem to be much agreement. Oaks are wind pollinated—they flower before the leaves are fully opened and produce copious amounts of easily blown pollen. Chestnuts, relatives of oaks, are often viewed as wind pollinated by farmers who grow them, with insects visiting the flowers as well. I believe the evidence suggests that chestnut is primarily insect pollinated. Like the oaks, chestnut flowers are borne on a long spike containing many male, pollen-producing flowers with a few female, seed-producing flowers at the base. Unlike the oaks, chestnuts produce flowers after the leaves are fully expanded. Foliage would block the flow of pollen from plant to plant, limiting pollination by wind. However, by flowering later in the season, conditions would be warm enough for insects to be out and active as pollinators. Furthermore, chestnut flowers, at least the Chinese and American hybrids in my home orchard, are showy with a strong, unpleasant scent that attracts a diversity of flies and other insects. There is no need to produce fragrance if the wind is the pollination vector. Chestnut pollen is also sticky—necessary for insect pollination but inhibitory to wind pollination. On the other hand, chestnut is reflected in the pollen record, like wind-pollinated trees, so some reproduction likely occurs by wind pollination as well.

The spiky husk that encloses the developing chestnut strongly indicates that small mammals were the major seed predators. While still developing and green, the nut is surrounded by a dense network of spines. At this stage, the fruit are like green sea urchins with hundreds of spines pointing in all directions. Thick leather gloves are needed to get into an immature chestnut fruit at this stage, but you still need to be careful. Despite such formidable defenses, squirrels break into the husks and consume chestnuts, though the spines slow them down. However, when the seeds are mature, the husk naturally splits open and releases the nuts to the forest floor, free and unprotected, to be consumed, hoarded, and hidden away.

Once seeds mature, those same seed predators that were the enemy now become great collaborators of dispersal. In contrast to fleshy-fruited species, chestnut seeds function as both the dispersal object and the disperser's reward. The combination of roles is a tricky situation to balance for the plant. If the seed predator is too successful, consuming a large portion of the year's crop, few seeds may remain to germinate in the spring. Chestnut trees must produce both a sufficient number of seeds to satiate predators while ensuring that enough remain in the spring potentially to grow into trees.

Dispersal is necessary for success for a variety of reasons. First, a new seedling that did not disperse would be deposited directly under the large, established tree that produced the seed. This location represents a competitive environment that is not likely to ensure success unless the maternal tree dies the year after seed production. Second, the soil around the maternal tree likely contains mycorrhizal fungi, beneficial fungi associated with plant roots, which could help the young tree grow. However, the soil probably also harbors pathogens that could be deadly to a newly emerged seedling. Those seeds that can germinate and grow will do so in high densities with their close kin—at least half-siblings. With broadly similar genetics, the way these seedling siblings will do everything will be similar, representing the worst-case scenario for competition and seedling success. The ability of seeds to be dispersed means they may colonize a new forest gap or push the local patch margins, expanding the boundary of where existing trees grow. In this way, a successful chestnut that disperses its seeds farther away can produce many more offspring than plants located in the population's center. The vast majority of seeds and seedlings always die, so this is a numbers game at heart. The more seeds a chestnut produces and disperses, the more likely a seedling will survive to maturity. Slight competitive benefits to seedling survival can form strong evolutionary pressures for dispersal over the generations.

It is interesting to think about the American chestnut fitting into the complex suite of interactions across species representing the forest food web of seeds, seed predators, and the seed predator's predators. Most acorns contain appreciable amounts of tannins, chemical defenses that bind with proteins to make them less digestible. Therefore, when consuming seeds with tannins, the seed predator gathers less protein from them, reducing predator growth and reproduction over time. There is variation among oak species in the tannin content of their acorns and variation within species. Seed predators can differentiate acorns based on their tannin contents when given a choice. In marked contrast to oak acorns, the chestnut is readily edible. With their lower tannin levels and much more palatable seeds, chestnuts would have been a prized food item.

As an illustration of the relative palatability of chestnuts as compared with acorns, I offer the following anecdote. While in graduate school, one of my lab mates was a gifted baker. He returned from a field site during an oak mast year with a large sack of white oak (*Q. alba*) acorns, considered a more edible oak species. Over the next several days, he shelled each acorn and soaked them in water, changing the water time after time to leach the tea-colored tannins from the nut pulp. What remained was a tiny mass of what he thought was suitably de-tannined carbohydrate. The cookies that he offered to the group proudly, we greeted enthusiastically, but we rapidly retreated to guarded politeness. The confection was like a walnut cookie that had included a few bags of tea leaves as an ingredient. After the lab meeting that day, I asked if he would bake with acorns again. His simple reply—"I don't think so."

In contrast, chestnuts can be simply roasted in the oven and their cooked seed eaten with no additional work necessary. The loss of such an easily digestible and abundant resource in forests must have massively affected the predators that fed on chestnuts and the wildlife that fed on seed predators. Surely, the American chestnut influenced myriad interactions when it was a major component of forests. We can only surmise the true impact of the tree's loss on forest ecosystems.

## The Blight

Chestnut blight was first noticed on American chestnut trees planted in the New York Zoological Garden when the trees mysteriously began to die in 1904. Initial treatment and eradication attempts were not successful. By

1910 the disease was also in New Jersey, eastern Pennsylvania, and Connecticut. A decade later, the disease was halfway across Pennsylvania and had progressed northward to Vermont's and New Hampshire's southern borders. An additional 10 years saw the disease throughout the northern extent of chestnut's range and extending south and west to a line that passed through eastern Ohio and central West Virginia to central North Carolina. At this point, the disease was in the heart of the chestnut's range, where the tree achieved its greatest stature and density. By 1940, the blight reached central Indiana, central Kentucky, central Tennessee, and Georgia's mountains. By 1950, nearly all chestnuts in North America were dead or infected, except for isolated populations at the margins of the range and chestnuts planted outside their range. This epidemic's body count was somewhere around four billion trees, though exact estimates are tough to generate.

The chestnut blight is not unique to North America; it also occurs on *Castanea sativa* in Europe. Roman colonists planted the European species for food and timber wherever they went, greatly expanding the species' range northward. Well after the blight was rampant in the southern Appalachians, it was discovered in Italy in 1938. European chestnuts are more resistant to the disease and succumb more slowly, but the ultimate results have mainly been the same. Chestnut blight continues to spread throughout Europe's chestnut populations.

What causes this deadly blight? It is an ascomycete fungus that infects woody stems' living tissues (Dutch elm disease is also an ascomycete, chapter 1). The fungus was first identified in 1906 as *Diaportha parasitica*, a species new to science. Further evaluation decided that it should be in a different genus, so the name changed to *Endothia parasitica*. In the late 1970s, the name changed yet again to *Cryphonectria parasitica*. Vigorous debate ensued, but this last name appears to have stuck.

Infection of a chestnut occurs when a fungal spore reaches a wound on a stem, and the fungus grows into the living tissues. As the fungus spreads out in all directions, it kills the cambium, a plant's growth tissue, and other living tissues, producing toxins as it goes. Eventually, the fungus grows completely around the stem, effectively girdling the tree and disconnecting the roots from photosynthesis in the canopy. When the cambium is damaged, any adjacent cambium cells that remain alive generate new stem sprouts from the bark. In larger stems with thick bark, this is often the first symptom to appear. Eventually, the bark will erupt into a canker, a sunken manifestation of the damage that has already occurred from the growth of the fungus. As the infection spreads, the damage increases, reducing growth and reproduction, and eventually killing the entire aboveground portion of the tree.

As with many pathogens, the fungus has two types of spores, matching two distinct reproductive modes. There are technical terms for each type of spore, but the key idea is that there are both sexually and asexually produced spores. When *Cryphonectria* hyphae are exposed to light within the canker, the fungus produces spores asexually. These spores are made at the tips of the hyphal threads, forming long chains of sticky, infectious propagules. Insects, birds, and mammals often visit open tree wounds seeking the sugars that ooze from them. After visiting a canker, the insect or vertebrate becomes a vector for the disease, spreading the spores from tree to tree, maybe even to another wounded stem. The distance traversed in this way is often short but is effective for disease spread within a chestnut population. Additionally, this is an effective and reliable way for the fungus to reproduce, as no mate is required.

When a fungus finds a compatible mate within the same tree, the hyphae can fuse, initiating the sexual cycle of the pathogen. As with Dutch elm disease, the sexual spores are produced within a sac—the ascus. However, in chestnut blight, the spores are forcibly discharged from the sacs into the air, allowing potentially long-distance dispersal on the wind. Together, these two modes of reproduction effectively and rapidly spread the infection across the forested landscape. When a blight infection first colonizes a chestnut stand, it is often by sexually produced spores that moved the distances between stands. Following the initial infection, sexual reproduction and the potential for further long-distance movement would be only as likely as two compatible fungal spores colonizing the same tree. This event might be reasonably possible in a tree near an infected population that was receiving many spores of many genotypes but much less so when moving long distances. Where infection rates are low, the disease can still spread asexually from tree to tree until one tree is colonized by two different fungi, finally allowing sexual reproduction to occur.

When most of the chestnut trees, the disease's host, are dead, we might similarly expect the *Cryphonectria parasitica* population also to decline. This decline should result in a dramatic decrease in the pathogen burden that any recovering or surviving chestnut faces, at least until the disease recolonizes the stand. We see this sort of episodic cycle between the Dutch elm disease and the elms (chapter 1), with cohorts of new seedlings being able to recover for a while, potentially even long enough to reproduce, and then the disease returns and kills them. However, chestnut blight is not just a disease of chestnuts; it can survive on other hosts. It isn't particularly pathogenic on other tree species and does not cause noticeable disease, so it may lurk undetected. The ability for the fungus to persist means that the former chestnut stands retain relatively high pathogenicity toward American chestnut long after the canopy individuals have been killed.

As chestnut blight spread throughout its host's range, it left behind a valuable commodity standing dead in the forest. Salvage logging occurred in many areas, and the rot resistance of the chestnut timber contributed to the economic success of this enterprise. However, the fungus can remain reproductive after the chestnut tree dies, and so this activity may have also helped transport the disease. From a human perspective, this disease also generated a glut of wood that became available in a relatively short period. Standing dead chestnut trees are not barren of life but often become the target for wood-boring beetles whose larvae consume wood. This pulse of food resources for the beetles likely also increased boring beetle abundance in forests where chestnut had been abundant. As there isn't much nutrition in wood, the beetle larvae must consume a lot of it, riddling the wood with their long holes. Wood from such trees became known as wormy chestnut and was widely used in furniture making and trim carpentry. The prevalence of beetle holes led some erringly to associate the beetle with the tree's death.

While I was in graduate school, friends of ours bought a house built as this pulse of chestnut wood was flooding local markets. This house was a monument to chestnut. All of the woodwork in it was chestnut—everything. The doors, the floors, the window trim, the staircase, the baseboards were all that rich brown, strong-grained wood. It was amazing. You can also glue any wood as a veneer on chestnut to cover up the holes and make beautiful furniture. My wife's family had a piano that was walnut burl veneer overlaid onto chestnut. As "repurposing" and "salvage lumber" has become of more interest since the early 2000s, chestnut timbers from old barns and industrial buildings have also come onto the market, nearly always sold as wormy chestnut.

Modern molecular techniques have finally confirmed the biogeography of chestnut blight's spread. The disease was always clearly Asian in origin, as species of chestnuts from Asia are largely blight resistant, having evolved alongside the pathogen. But where in Asia? Researchers from China, Japan, and the United States combined efforts and did a comprehensive sampling of chestnut blight fungi across its native and invaded ranges. There were genetic differences across the Asian, European, and North American populations. Differentiation would be expected as the tree species infected would be different and lead to genetic changes over time. The greatest genetic diversity of chestnut blight was found in China. Again, this is not surprising as the geographic area involved is quite large and there are several *Castanea* species native to that region. However, the data revealed the strongest genetic affinities between native Japanese populations and Europe's and North America's invading populations. This pattern neatly fits with early plant importation records for New York and the similarity of the disease in Europe and North America. Knowing how the disease spread doesn't change how we treat the blight or generate any other practical outcomes, but the story is complete now and serves as a warning about the uncareful transport of species.

As the blight spread, some trees began to recover and heal their cankers. Instead of developing the normal sunken appearance of chestnut blight cankers, the cankers on recovering trees became swollen and enlarged. Most importantly, the trees did not die. Similar observations were made in Europe on their infected trees. The cankers that healed became referred to as hypovirulent cankers—a disease with a lowered virulence. It turns out that the *Cryphonectria parasitica* fungi in cankers that were healing had become infected with a pathogen of their own, one of several viruses. For a time, hypovirulence appeared as the holy grail of chestnut's salvation. Researchers began growing the fungus with the virus and applied it to cankers in the field with success.

Based on this early work, biological control programs were developed with great anticipation. The reality, however, was much less clear. For any biological control to work correctly, the biological control agent must establish a population and spread with its target. This is where things fell apart for hypovirulence. European chestnuts responded better to biological control than did American chestnuts, which isn't surprising as they are less susceptible to the disease. Trees growing in an orchard or other open conditions responded better than plants growing in higher densities or forests. Places infected with a lower diversity of the blight fungus responded better to biological control than sites with greater pathogen diversity. While hypovirulence was an exciting idea and a fascinating case study of a tree–fungus–virus interaction, the ability to control chestnut blight in natural forests appears relatively small. Recovery must depend on other mechanisms.

After the blight, relict populations of American chestnut remained in locations along the periphery of the original distribution and areas where the trees were transplanted. One of these populations was in West Salem, Wisconsin. This population, established by planting a few individuals well outside the tree's natural range, was therefore well isolated from chestnut blight. Chestnut persisted in this forest with little apparent regeneration for decades. In the 1970s, cattle were excluded from the forest, followed by some logging, resulting in a dramatic canopy opening. Together, these two factors allowed chestnut seedlings to survive and capitalize on the higher available light. This ecological release resulted in a pulse of recruitment for chestnut and an extended period of growth that dwarfed the growth of other trees in that forest. Some of these trees exceeded 1 foot (0.3 m) in diameter by 2001, when cores were collected to determine the tree's age and growth rates. Not only does the West Salem site affirm the robust physiology of American chestnut suggested by Latham's work, but it also gives hope that the species could rebound if the blight could be successfully combatted. Alas, the dreaded blight arrived in West Salem in the late 1980s, decimating the population.

## The Blight's Aftermath

When chestnut disappeared from Eastern North America's forest canopy, it left resources underutilized in a large swath of forest environments. On the drier end of its environmental range, several species of oaks and hickories (*Carya*) replaced chestnut. This transition was famously predicted by Catherine Keever, an early female plant ecologist, only a few years after the blight and confirmed a few decades later. By the 2020s, oaks that initially replaced the chestnuts range from approximately 70 years to over 100 years old in the earliest infected forest stands. These trees are now mature and will begin to replace themselves in the forest. It is interesting to think that the contemporary abundance of oaks can be traced back to a disease in another tree species that led to a large, almost identically aged cohort of oak recruits that form the forest canopies of today.

On moister sites, a different suite of species replaced the chestnut. Here, neighboring species such as maples, tulip poplar, and cherries (*Prunus serotina*) expanded to fill the void in the canopy. The ecologically fascinating thing about chestnuts was that, regardless of where they grew, more than one species typically replaced them when the chestnut succumbed to the blight. This is yet another testament to the broad environmental tolerances of chestnut. Perhaps it was only the rapid opening of the canopy that allowed multiple tree species to expand at once. Eventually one species may dominate the forest canopy over the next few generations of trees; only time will tell.

While chestnut has essentially been eliminated from the canopy, it persists in many forests. The ability of chestnut to resprout prolifically following cutting is one reason that chestnut was often abundant in forests following logging. Similarly, when the chestnut blight kills the chestnut tree, the roots remain alive and often generate several sprouts from remaining energy reserves. This ability to resprout has effectively created a population of sprout-shrubs that persist and grow in the forest understory. The only way I have ever seen a true American chestnut is as a set of sprouts on an old stump. I almost always mistake them for a chestnut oak (*Quercus prinus*) because of the similarity in their leaves and habitat. Long-suppressed chestnut seedlings also remain in the forest understory in some forests. When given an opening, understory plants can shoot into the forest canopy. And, if a suitable mate is available for cross-pollination, these trees may reproduce a few times before the blight catches up with them. The young chestnuts, or young sprouts on old roots, appear to be somewhat protected from infection for a short while.

In areas where the chestnut was abundant, sprouts can represent a large amount of biomass in the forest; therefore, the chestnut maintains a minimal shadow of its former glory. The number of chestnut stems can be quite large in these forests, and they likely still retain some function in resource uptake, photosynthesis, and as food for herbivores. Some have argued that this new life-form for the chestnut will be its new ecological niche, living in perpetuity as an understory plant. This idea simply ignores reality. The American chestnut is functionally extinct across the majority of North America. As generation after generation of chestnut sprouts develop, get infected, and die back to the roots, individuals will be lost. Each stem that dies is another chance for an opportunistic fungus or another pathogen to infect the root. Without effective reproduction, the species faces the slow, inevitable decline toward extinction.

Another largely ignored consequence of the chestnut blight has been the decimation of a closely related species, the Allegheny chinquapin (*Castanea pumila*). This shrubby species grew throughout the southern and central portion of the American chestnut's range and extended into the southern coastal plain. Chinquapins are very similar to the chestnut, producing smaller "chestnuts" in tiny spiky burs, just like a chestnut. The size of the nut is more similar to that of a wild hazelnut (*Corylus*). There are related species in California and Oregon (*Chrysolepis sempervirens* and *C. chrysophylla*) that produce what appear to be nearly identical fruit to their East Coast counterpart. These species have escaped the blight to date but could succumb if the disease is transported westward. Let's hope we never find out if those species are susceptible to the blight! Unfortunately, there isn't much we can say about the loss of the chinquapin. Ecologically, we know very little about the role that shrubs play except for a few wildly invasive species (chapter 7). For whatever reason, people just don't study shrubs all that much.

## Restoration—Breeding and Distribution

Because of its charismatic position in forests and American lore, American chestnut has long captured the interest of those who would like to see it return to a place of prominence within its former range. Quite early into the blight epidemic, breeding programs were started in the 1920s that crossed American chestnut with Chinese chestnut (*C. mollissima*) or other Asian species as an attempt to introduce disease resistance. While these first attempts used tried and true methods, breeding trees, even those that reproduce when young, is a slow endeavor. Early attempts were largely unsuccessful, and they were abandoned in the 1960s, though their orchards remained. Starting in the 1980s, a corn geneticist, Charles Burnham, restarted the breeding program and was a founding member of the

American Chestnut Foundation. This foundation has led the way to save the American chestnut and has brought us to the cusp of reintroducing chestnut into forests.

Their breeding plan is quite simple, starting with the plants established in the prior breeding programs. Each generation is crossed back with American chestnut pollen, effectively reducing the non-American genetic component by half each time. The resulting seedlings are grown for several years, then inoculated with the blight, evaluated for resistance, and only the best selected to pass on to the next stage. This selection must be very discriminating—the stronger the selection pressure, the faster the desired genetic changes will occur. Of course, genetic diversity is critical for the continued success of a species, so multiple hybrid chestnut lines are produced, each bred with multiple American chestnuts. Each stage of the process involves generating tens of thousands of seedlings, growing them for a few years, culling to a smaller number by screening for resistance, then moving forward once they reach reproductive maturity. Each seedling's parentage must be tracked meticulously. The ultimate target is to produce not a single resistant line but an orchard full of trees with genetic resistance to the blight. These trees must also be predominately American chestnut from a morphological and ecological perspective. From these orchards, the hope is to repopulate the landscape with trees possessing the genetic composition to persist in the disease's face.

With such an extensive geographic range, covering a variety of climatic conditions, producing a single, though diverse, chestnut line may not encompass the ecological amplitude necessary for successful regeneration everywhere. Therefore, this large-scale breeding program has been replicated focusing on local conditions in many regions. This truly arduous task has been conducted by state chapters of the American Chestnut Foundation. Several of these seed orchards are scheduled to complete their selection process between 2020 and 2030, with seeds then distributed throughout chestnut's former range. On a trial basis, seedlings are being planted in various locations to verify the success of the breeding for disease resistance, to develop planting protocols, and to assess potential threats to the new plantings.

Another, more modern, approach has been employed by a team of researchers based at State University of New York's College of Environmental Science and Forestry. Their approach has been to transform the

American chestnut with genes that will confer blight resistance. The first step is to get chestnut cells from embryos into tissue culture. These cells have not differentiated yet and so can be used essentially as stem cells. The gene or genes of interest can be inserted into the DNA of the chestnut cultures. Once this is completed, the tissue culture can be grown in perpetuity, making generation after generation of genetically transformed plants. Of course, all plants from that single tissue culture would be genetically identical, so other cultures started from other seeds would be needed to generate genetic diversity. Once transformed, the new genes can be transferred to other plants through standard plant breeding, passed on in both the pollen and eggs necessary to form seeds.

So far, there have been two genetic modification approaches attempted. One has placed the gene for oxalate oxidase into the chestnut. Oxalate is the primary toxin produced by *Cryphonectria parasitica*, so being able to degrade it prevents disease symptoms. The viruses that cause hypovirulence also reduce oxalate production, so the transgenic plants mimic a proven disease management mechanism. The other approach has been to isolate genes for blight resistance in other chestnut species and introduce them into American chestnut cell lines. While this entire process may seem overly complicated, tissue culture is widely employed in propagating plants from orchids to forestry trees. The clear benefit to this approach is that the plants produced will be almost entirely American chestnut except for one or a few genes, and the results can come about quickly.

While I whole-heartedly support and anticipate chestnut restoration, it will likely generate some vigorous debates as we move closer to planting trees in a restoration setting. Is it truly acceptable to plant trees that are some percentage a non-native species with the intent of doing so across a large region? Are there other genes that accompany the disease resistance that will present management problems decades down the road? Similarly, we often are skeptical of genetically engineered organisms. Though they will need permitting to proceed, is society going to accept transgenic trees in our landscape? There are famous cases of genetically engineered crop traits escaping into closely related wild species. While in the case of American chestnut I cannot think of any plants other than the chinquapin to which such gene flow could occur, this is a concern commonly voiced. It has always been fascinating to me that we may effectively be transferring the same genes, some by traditional plant breeding, some by genetic transformation, but we never worry about the plants produced using traditional breeding methods.

We see similar debates in nearly all aspects of society. Every new approach has its detractors and supporters. Take the energy sector—fossil fuels increase carbon dioxide and produce other pollutants. Critics argue that solar arrays work only when the sun is out; turbines kill migrating birds; nuclear reactors generate waste and can be dangerous; hydroelectric facilities alter hydrology and fragment rivers. There is simply no pleasing everyone or perhaps anyone. Despite the philosophical debates that will almost certainly arise, the pending restoration of chestnut is a miracle of modern ingenuity, collaboration, and largely private financial resources. The success of the American Chestnut Foundation and its collaborators is a testament to what can be achieved. Their approach should make us much more optimistic as we proceed, tree by tree.

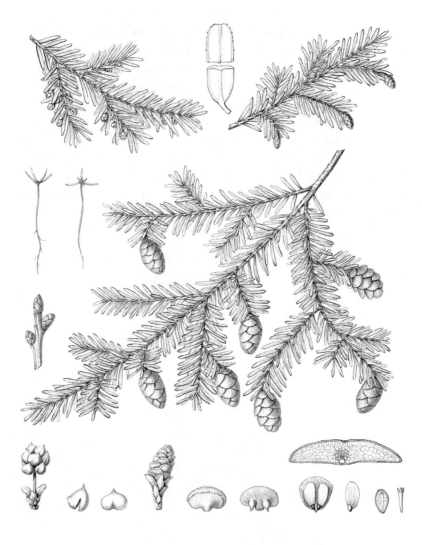

TSUGA CANADENSIS, Carr.

Leaves, cones, seeds, and seedlings of eastern hemlock,
*Tsuga canadensis.*

# 3

# EASTERN HEMLOCK—*TSUGA CANADENSIS*

In 2011, my wife and I accomplished a long-term dream of moving to the country. We purchased an 80-acre (32-ha) farm, complete with three barns in various states of disrepair, two garages, an old smokehouse, two ponds, and the original farmhouse that we likely would have been better off bulldozing than remodeling. Not to be deterred, we bought the farm, literally, and began the several-year renovation process. While most of the work focused on the farmhouse, a disaster in its own right, I also started clearing the outbuildings of the accumulation that occurs only when you have no space limitations. I was disheartened when I finished junk removal because the wooden barn was in worse shape than I had initially thought.

My father and I addressed the obvious framing deficiencies, and I set about tearing out and replacing the hayloft floor. The roof was the next task at hand, as a complete floor in the hayloft revealed precisely how badly the roof was leaking. At about this time, the local paper had an ad for someone who could "make your old barn new again." After making an inquiry, I was put in touch with Melvin, an Amish man. Melvin had a straightforward way of speaking with a subtly Germanic accent, and he was extraordinarily knowledgeable about barn construction.

While I had him on site looking at the roof, I asked about residing the barn. I told him that I liked wooden barns and was hoping to maintain that and restore the barn—we bonded immediately. The barn was clad in failing tongue-and-groove pine boards that retained the barest vestige of traditional barn-red paint. Melvin quickly eliminated the idea of pine, as the knots of the open-grown pine available today are too large and will pop out as the wood dries; he knew he would be back for years to come replacing failed boards. I suggested oak board-and-batten siding from a local lumber mill as a more responsible choice. Melvin equally discarded this idea out of hand as the material would warp as it dried, pulling away from the building—this time with hands illustrating the warping action of the wood. He was getting more animated and pointed directly at me now as he said, "What you need is hemlock."

This farm is in east-central Illinois, with the nearest appreciable hemlock stands in northern Wisconsin or eastern Ohio—this was not a wood commonly available in the area. Melvin told an arm-waving story of a buddy in western Pennsylvania with a sawmill and the Mennonite trucker that went "visiting" Amish communities, delivering materials from area to area. We shook hands, and he left to calculate a price for the roof and place a call about the cost of hemlock. The results were spectacular. I would love to say that I possess the only hemlock barn in the area, but Melvin was so happy with the results he sent a family by our place the following spring to look at our barn before Melvin remodeled theirs.

## Forests of Hemlock

The distribution of hemlock is narrowly constrained to mountainous regions in the southern Appalachians but much more broadly distributed in New England, northern Michigan, Wisconsin, and adjacent areas of Canada. In the southern Appalachians, hemlock is restricted to higher altitudes and the deep valleys that channel cold, moist air down the

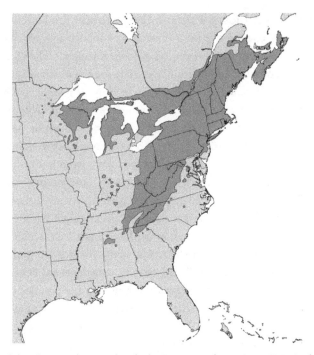

Natural distribution of eastern hemlock, *Tsuga canadensis*. From U.S. Geological Survey Geosciences and Environmental Change Science Center.

mountainsides. Northward, hemlock moves down in elevation, eventually becoming common throughout the landscape but never in boggy or saturated soils. Often, eastern hemlock develops into almost pure stands, casting a deep shade that primarily allows other hemlocks to regenerate.

Regeneration of hemlock is interesting. Following a fire or other disturbance that exposes mineral soil, germination can lead to a large, relatively even-aged stand of trees. However, in areas where the thick litter layer is intact, the species preferentially regenerates on rotting logs and stumps. This regeneration niche leads to spatial patterns of mature hemlocks that persist long after the log has completely decayed and disappeared. Often you can find good-sized hemlocks growing in a line representing where they had all established on the same fallen tree. In a Pacific Northwest old-growth forest, I found several massive western hemlock (*T. hetero-phylla*) trees perched above the soil surface, supported by extensive root systems that had centuries prior grown down through a fallen log.

As an incredibly shade-tolerant tree, hemlocks exhibit a classic slow and steady growth habit. A seedling may be only an inch or two in height in its first year, with an equally minuscule root system. This small size is why the species depends on continually moist, but not wet, conditions for seedling establishment. It also explains the species intolerance to trampling. Seedling growth is ruinously slow. Even under ideal nursery conditions, a 3-year-old tree seedling may only be between 5 to 9 inches (13–23 cm) tall—trivial compared with nearly any other tree species' growth. Seedlings of many trees exceed that height within their first growing season, even when grown under field conditions. Young hemlock seedlings typically depend on sheltering shade for growth and survival, doing well under young established trees of other species. Forestry practices to encourage hemlock regeneration hinge on the practice of shelter-wood cutting, leaving some protecting cover to keep the young seedlings from drying out. From there, the growth of shade-suppressed seedlings can still be relatively slow, persisting as seedlings for 100 years or more, even multiple centuries in some cases. Hemlock seedlings respond to canopy openings with increased growth. However, too much exposure reduces photosynthesis as they get stressed in the drying heat of a high-light environment.

Growth rates in hemlock reflects one of the great trade-offs in ecology. The greater tolerance a species has to low levels of a resource, light in this instance, the lower the plant's ability to capitalize on that resource when it is abundant. So, while eastern hemlock is one of the most shade-tolerant species, surviving on as low as 5% of ambient light or less, that tolerance also constrains it to be a remarkably slow-growing tree. Slow and steady wins the race, so these trees reach massive sizes. Mature stands typically have trees around 400 years old, 3 feet (1 m) or more in diameter, and well over 100 feet (30 m) tall.

I have walked through mature hemlock stands in the Shenandoah National Forest while hiking the Appalachian Trail. No words can capture how those forests feel other than to say they are a cathedral of trees. The deep litter layer that accumulates under hemlock stands keeps out under-story herbs and shrubs, muffling your footsteps to maintain a church-like quiet. The tree height is immense, with no subcanopy trees obscuring your view of the evergreen canopy above, supported by the trunks that encompass you like gothic columns. It is also dim as the canopy filters out nearly all light, often leaving only a percent or two of solar radiation to strike the forest floor. Most hemlock stands occur as distinct patches; within a few moments you emerge, squinting, into the crunchy-leaved, multilayered forest that dominates the surrounding area.

With such an extensive geographic range, hemlock grows with a great many other tree species. Lucy Braun includes eastern hemlock in the mixed mesophytic forest at the southern end of its range, grading into the hemlock–white pine–northern hardwoods at the northern extent. Given hemlock forms localized, monospecific stands, it is often a minor component of the diverse mixed mesophytic forests but can be a canopy dominant in cove forests along streams and steeper, protected slopes. As southern forests transition to more northern deciduous forests of Pennsylvania and New Jersey, this pattern is maintained. Oaks (*Quercus* spp.), maples (*Acer* spp.), hickories (*Carya* spp.), and tulip poplar (*Liriodendron tulipifera*) form the bulk of the surrounding forest matrix that includes dense hemlock stands. Farther north, eastern hemlock is an important component of the white pine–hemlock–northern hardwoods forest of Michigan, Wisconsin, and New England. There, hemlock grows in association with birch (*Betula lenta*), American beech (*Fagus grandifolia*), sugar maple (*Acer saccharum*), and, of course, white pine (*Pinus strobus*). Here, hemlock can occur in monospecific stands along ravines and slopes where it commonly dominates, or it can be part of a mixed forest. There is a lovely old-growth white pine–hemlock forest in northern Pennsylvania, with massive examples of both trees growing in nearly identical abundance. At its northern extreme, eastern hemlock ends abruptly at the boundary with Canada's boreal forest.

Often ignored, there is an additional hemlock species in the eastern United States, the Carolina hemlock, *T. caroliniana*. This species occurs in isolated populations on rocky slopes of the southern Appalachian highlands, encompassing extreme northern Georgia, eastern Tennessee, and western portions of South Carolina, North Carolina, and Virginia. The bulk of locations occur in North Carolina. The species reaches 100 feet tall and has slightly longer needles that come off the stem at all angles, rather than flattened into a single plane as in eastern hemlock.

Carolina hemlock also produces larger cones than its more abundant relative. Because of its rarity, its economic impact is mainly as an ornamental tree.

Hemlock wood has been used primarily for rough construction, railroad ties, siding farm buildings, and, to some extent, as pulp for the paper industry. Paper production from such a slow-growing tree seems a shame, considering the much faster-growing trees that typically surround hemlock. Hemlock's wood fibers are too short to produce high-quality paper, giving *Tsuga* an even lower value for paper production. The wood of eastern hemlock tends to split along its growth rings, a flaw known as ring shakes, making it useless for structural timbers or fine construction purposes. Hemlock bark, like American chestnut, was also formerly a critical source of tannins for processing animal hides. With the widespread insect outbreak decimating eastern hemlock populations, logging has attempted to keep pace before the wood resource is lost. Logging may help slow the invasive insect's spread in the pest's inexorable rush to cover all hemlock like a biblical plague. The landscape ultimately ends up in the same state—sadly devoid of hemlock.

## The Invasion of the Hemlock Woolly Adelgid

In the spring of 2002, I was in the Great Smoky Mountains National Park, helping to lead a group of college students on a post-graduation hiking trip. My colleague, Andy Methven, led the outing, since he had done his doctorate in Knoxville and knew the Smokies quite well. Each day, we would drop the students off, instruct them clearly on the day's hike, then proceed to drive to the pick-up spot, leave one van, then return to the drop-off site with the other van to start our hike. Andy and I were left to walk alone, chatting and viewing the abundant spring wildflowers. Andy is a mycologist and felt compelled to stop at every log, mushroom, or other fungal habitat to see what was there. He was a constant font of fungal names and trivia, which largely passed over me with little retention.

There was one exception. *Ganoderma tsugae*, the "hemlock varnish shelf," is a bracket or shelf fungus that grows primarily on living and dead hemlock. It is a pretty fungus with a shiny reddish-brown top that often fades to a creamy white toward the bracket's margins. Charismatic, as far as fungi go. Even more interesting is the specialist beetle, *Megalodacne heros*, which predominately feeds on *Ganoderma tsugae*. This attractive beetle, large at 5/8 inch (1.6 cm), has bold orange and black bands across its back. Its bright colors warn away any would-be predators, signaling the beetle's unappealing taste.

Days later, we poured exhausted students into the vans and started the journey back to Illinois. To stay awake, I tuned the radio to the local National Public Radio station. The first report that morning was that the hemlock woolly adelgid had been found in the Great Smoky Mountains' southern end. As the non-native insect moved from established populations in the north, all the park's hemlocks were now threatened. I had seen how fast this invader killed hemlock trees when I was in New Jersey, so I knew what was in store for this national park. While the hemlocks were likely doomed, that pretty fungus and even prettier beetle would do well for a while, thriving off the pulse of dead and dying hemlocks. Ultimately, they would fare no better, restricted to whatever remaining trees could support them.

Adelgids are plant-feeding insects, commonly known as "spruce aphids" or "pine aphids" and similar to the true aphids (another insect family), that often specialize on tree genera. Often, adelgids are simply referred to as aphid-like. The offending insect on eastern hemlock is *Adelges tsugae*, with the species named for its target tree genus, *Tsuga*. This insect species was first described in Asia but then was found in the western United States in the early 1920s and the eastern United States in 1951.

All of the North American hemlock woolly adelgid populations were initially thought to originate from Asia, but there is some doubt about the origins of the western populations. The hemlock trees there (*T. heterophylla* and *T. mertensiana*) have much greater resistance to the adelgid, with mortality caused in only severely stressed trees. Furthermore, there is a native beetle in western North America, *Laricobius nigrinus*, that feeds extensively on the adelgid. There is also greater genetic diversity in western adelgid populations, suggesting either much greater introduction pressure through repeated introductions from multiple sources or a much longer history in the area. Together, these patterns could reflect ancient biogeographic processes by which hemlocks, their enemies, and their enemies' enemies were separated as the continents moved. Eastern hemlock may have lost its herbivore and, subsequently, resistance to its herbivore along their evolutionary odyssey.

In sharp contrast to western North America, eastern populations of the hemlock woolly adelgid are clearly introduced. The eastern hemlock has little to no defense against the herbivore, there are no native predators on the adelgid, and the populations of adelgids have little genetic variation. Genetic surveys have identified the source of the Eastern North American invasion as being from southern Japan. While the insects were known to be in the east since the 1950s, little happened until the mid-1980s, when adelgid populations in Connecticut and New Jersey expanded, and hemlock trees began to die. The hemlock woolly adelgid feeds on xylem ray parenchyma, living tissues associated with water-conducting cells, at the base of leaves or young stems. The adelgids also appear to inject toxins into the plant tissue. The insect's feeding impairs water transport, consumes plant resources, and causes leaf and stem mortality.

*Adelges tsugae* has an unusual life cycle compared with many insects, but it has worked out quite well for the species. First, hemlock woolly adelgids in North America are all females, producing eggs parthenogenetically, in other words, asexually. Overwintering adults lay eggs in fluffy, white masses in March and April. When the larvae crawl out, they can disperse on the wind or settle down to feed at the base of a leaf. Over the next few weeks, they feed, grow, and molt three times, becoming adult females sometime in June. This is the point at which things get strange. Some adults will be wingless and stay on the hemlock. These insects will lay eggs that produce larvae that will lie dormant until the fall, when they resume feeding and start the cycle all over again. Other adults will be winged and will fly off to find a spruce (*Picea* spp.), the species' definitive host—the host on which it sexually reproduces. The odd part of the story is that none of the spruce species in North America appear adequate for the insect to complete the sexual portion of its life cycle. Therefore, all the hemlock woolly adelgids in North America are asexually produced clones. Host switching and brutally effective asexual reproduction are common features in the life cycles of many parasites and pathogens, regardless of how odd it sounds. This invasion would be even more devastating if the adelgids could sexually reproduce, generating genetic variation. They would also have the potential to harm one or more North American spruce species.

While sexual reproduction is the primary mechanism for generating genetic diversity, this does not mean that the hemlock woolly adelgid lacks variation. Asexual organisms, with no opportunity for sexual recombination of genes, often have high mutation rates as a way of compensating. If mutations happen during egg production, then the new individual produced is different genetically, perhaps in a vitally important way. Cold is one of the primary limitations to the hemlock woolly adelgid survival in the north, often leading to significant winter mortality before the spring bout of egg laying. However, researchers have found considerable variation in cold tolerance among regional populations, with greater tolerance in northern insect populations. Such variation indicates evolutionary responses are occurring in this asexually reproducing species, helping it adapt to local conditions.

As flying adelgid individuals are reproductive dead ends, there must be a mechanism that allows the adelgid to spread among regional hemlock populations. The small size of the insects (5/16 inch [0.8 cm], not including the waxy threads, the "wool," covering its body) allows eggs and adults to spread by wind more than half a mile. Storms assist in passively moving adelgids, with hurricanes and other significant weather events linked to even greater dispersal distances. Wind, however constant, cannot direct dispersal to a host tree. However, adelgids are also carried by birds and mammals, particularly those frequenting hemlock forests, effectively targeting even the most isolated hemlock. Human activity as well has undoubtedly aided the movement of adelgids—from campground to campground, with logging equipment, or anywhere we move materials.

## Changes in Forest Structure

The losses of American elm and American chestnut predated modern scientific approaches that characterize contemporary ecological research, both from analytical and technological perspectives. While studies of previous tree losses were primarily descriptive, the loss of eastern hemlock coincided with advances that have given us a much better picture of the species' decline and its implications. Our research approaches now allow detection of subtle patterns in spread and effects that generate a detailed, mechanistic perspective on what is occurring.

When hemlock woolly adelgids colonize a patch of eastern hemlock, they may go initially undetected. Still, the insect population increases rapidly and can kill a tree in a year, though it often takes longer. Stressed trees are typically the first to succumb to the insect, especially suppressed trees—those beneath older, taller hemlocks. Stands on drier sites or with high densities of smaller hemlocks experience mortality rapidly, sometimes being destroyed in as little as 4 years. This observation has led to management recommendations to reduce stress in uninfected stands by thinning the hemlocks, removing subcanopy individuals.

One characteristic that lends a much-needed landscape perspective to the hemlock woolly adelgid outbreak is that the host tree is evergreen. While this is not all that unique, it does provide a way to document tree loss not possible for deciduous trees. Researchers can quantify hemlock loss remotely by examining winter aerial photography or satellite imagery when hemlocks stand out. Though there are pines and other evergreens in many of the affected forests, changes in green winter cover should be almost entirely related to the loss of eastern hemlock, particularly when the changes are rapid. Satellite imagery provides an even greater sensitivity level, detecting changes in the level of greenness in photosynthetic tissues that occur before tree death. Remote sensing technology has exploded rapidly, allowing assessments to include ever-increasing levels of detail.

Digital maps of changing eastern hemlock cover can be combined with other maps that include features such as slope, steepness, the direction a slope faces, soil type, size of hemlock stands, and distance to the nearest adjacent hemlock stand. Each piece of information is processed as a "layer" in a GIS (geographic information system) map and is used to understand hemlock decline. The results can be visually shocking, particularly when done at regional spatial scales. These analyses confirm that trees on drier sites, ridges, and slopes that face south or west succumb to hemlock woolly adelgids more quickly. There is also some indication that hemlock decline rates are more rapid in the southern as compared with the northern end of hemlock's range. Winter mortality for the adelgid is lower in the south, leading to more rapid adelgid population growth and more rapid tree impacts. The single greatest contribution of these studies has been to document the depth of the problem—although some places in the landscape are slower to be colonized by adelgids or take longer for the trees to die, nothing is sufficient to protect hemlocks. The maps tell the tale; if this adelgid pest is left unchecked, no hemlocks will escape.

The loss of eastern hemlock following an adelgid outbreak is function-ally permanent for that stand at a human time scale. Unlike the American chestnut, hemlocks do not resprout, necessitating that all regeneration be achieved via seeds. Unlike the American elm, young trees cannot avoid the insect pest long enough to reproduce. The narrow establishment require-ments of hemlock mean that there often are not hemlock seedlings wait-ing to replace canopy and subcanopy individuals as they die. A hemlock woolly adelgid outbreak usually involves hemlock seedling mortality, directly from the insects and indirectly from the physiological stress of rapidly increasing light. Surviving seedlings may take many decades to reach a reproductive size and will remain susceptible to hemlock woolly adelgids as long as the insects persist. Finally, hemlock seeds do not sur-vive for more than a few years in forest soil, restricting any regeneration to the period immediately following canopy tree death.

With their specific habitat requirements and deep shade tolerance, hemlocks typically occur in forest stands where most canopy trees are hemlock. Therefore, hemlock loss does not generate isolated forest gaps able to be quickly filled in by one or two trees. Instead, hemlock death is typified by large, continuous openings in the forest canopy. These open-ings represent massive opportunities for trees and other plants to colonize, including invasive plant species (chapter 7). Birches are one of the more opportunistic tree species that colonize following hemlock collapses, par-ticularly *Betula lenta*, the black or sweet birch. Birches do not regenerate under shade, so as their canopy individuals die, they will be replaced by tree species that are more tolerant of shade—maples, oaks, and American beech in many areas. As with the other tree losses, the cascading ripples of forest dynamics generated by the loss of hemlock will likely be evident for centuries as even-aged cohorts of trees come and go.

In the southern Appalachians, eastern hemlock often grows in associa-tion with a large, understory shrub, *Rhododendron maximum*. This ever-green shrub often forms dense thickets, 10–30 feet (3–9 m) high, under which only hemlock seems able to regenerate. As with all understory shrubs, they persist in low light but capitalize on canopy openings with rapid growth and reproduction. This shrub has been of interest to forest-ers for years as it rapidly spreads after logging and reduces tree growth and regeneration. Forestry practices often advocate the removal of this *Rhododendron* to ensure adequate forest regrowth. Suppose this species replaces hemlock on even a subset of sites. There, the hemlock woolly adelgid invasion would convert tall forests containing massive amounts of biomass to shrublands containing the merest fraction of that original biomass. Eventually, these shrubs will die and allow the regeneration of trees, but this could take an exceedingly long time. Foresters may need to remove *Rhododendron* to allow sufficient tree growth to regain any timber value in post-adelgid forests as well as after logging.

## Changes in Ecosystem Function

A significant paradigm shift in ecology over the last half century has been the advent of the ecosystem concept and ecosystem ecology. Unfortunately, people who love to latch onto scientific buzzwords have co-opted the word ecosystem and muddied its usage beyond all recognition. I find it clearer to define the ecosystem concept by the type of processes that it encompasses—the movement of energy and materials (nutrients such as carbon, nitrogen, or calcium) through a system. An ecosystem necessarily involves two distinct components, pools and fluxes. A pool is where energy and materials are contained—the carbon in living plants, the nitrogen in a pond, or the calcium content of soil. A flux is a transition, a movement, from one pool to the next—the release of carbon dioxide to the atmosphere from plant decomposition by soil microbes, the uptake of soil calcium by plant roots, or the loss of nitrogen in a pond through outflow. You cannot talk about pools without fluxes and vice versa. A rapid flux into a pool and a slow flux out will result in a large pool; a slow flux into a pool and a rapid flux out will result in a small pool. Knowing the size of pools without understanding the fluxes is tricky—a small pool may represent an unimportant feature to the system or a critical component that processes quite quickly. The dynamics of both are key to understanding an ecosystem.

Decomposition is a critical ecosystem process that varies greatly with the quality of the material that is decomposing. The most important quality of leaf and woody debris that determines decomposition rate is the ratio of carbon to nitrogen. Plant tissues always contain more carbon than nitrogen because the physical structure of plant tissues is primarily composed of carbon. With dramatically more carbon per nitrogen molecule, microbial processing will be limited by nitrogen availability. Nitrogen forms the functional core of microbially produced enzymes that break organic matter down and forms the proteins necessary to make more microbes. The carbon content of leaf litter, in contrast, is the material respired by microbes to produce energy. As nitrogen contents increase, the rate of microbial decomposition will also be quicker. Balancing nitrogen and carbon content is vital to backyard composting as well—too much brown carbon and composting occurs slowly; too much green, nitrogen-rich material, and the compost will process too quickly, go anaerobic, and stink. Eastern hemlock leaf litter, because of its physical structure and carbon-based defensive chemistry, is very high in carbon relative to nitrogen and decomposes very slowly, building up a thick pool on the forest floor.

As hemlock woolly adelgids move into a hemlock stand, their feeding begins to open the canopy, allowing more sunlight and rainfall to reach the forest floor. Temperature and moisture also regulate decomposition, so warmer and wetter litter supports higher microbial activity and decomposes more rapidly, increasing flux out of the litter pool. Furthermore, adelgid feeding changes hemlock leaf chemistry by increasing nitrogen content, resulting in improved litter quality for microbial decomposition. Some foliar nitrogen dissolves in rainfall as it passes through infected hemlock canopies, fertilizing the litter below and additionally increasing decomposition rates. Litter decomposition releases nitrogen available for plant uptake. As hemlocks suffering from woolly adelgids are often dead or dying, nutrient uptake may be limited, resulting in unused nitrogen leaching into streams or groundwater, lost to the local forest. This flux into stream waters may cause problems in downstream habitats that receive the nutrient inputs. Over time, the remaining hemlock tissues decompose, maybe taking a decade or more to work through the large pool of recalcitrant litter. Therefore, the adelgid's ecosystem impacts are likely to last long after the insect and trees have disappeared and to spread far beyond the trees' original area.

The tree poised to benefit immediately from hemlock decline is again the birch, *Betula lenta*. At least part of this tree's success following hemlock decline may be due to higher nitrogen availability in declining hemlock stands. Birches do very well with higher nitrogen, allowing them to outcompete other tree species in the new habitat and rapidly develop thick populations of trees. As the decompositional legacy of eastern hemlock litter wanes over time, the birch's litter chemistry will come to dominate the ecosystem. Birches have much higher litter quality as there is much more nitrogen relative to the leaf carbon content. Therefore, *Betula* litter will decompose much more rapidly than hemlock litter, producing a much less substantial pool of decomposing leaf litter. The litter chemistry of birch will likely dominate former hemlock stands for decades until the birches themselves become replaced by shade-tolerant trees and their leaf chemistry. The forest's ecosystem dynamics will ultimately depend on which tree species replace the birch when they can no longer reproduce.

## Impacts of Hemlock Loss on Animal Populations

One of my favorite hikes that I have repeated in locations from the southern Appalachians of northern Georgia to northern Maine, is a steep walk along a hemlock-shrouded stream. The water rushes past, with the occasional flash of silver as a trout darts to snatch a tasty invertebrate that the flow brings its way, only to dart back to the lower flow that extends like a shadow beyond each boulder. The water is also bracingly cold—an aspect of their habitat upon which the trout depend. The shade provided by dense hemlocks is a characteristic of hemlock forests across this tree's entire geographic range. By intercepting light, eastern hemlock keeps these streams cold and clear of algae that would grow in sunnier locations. Even where the stream is too broad to be covered entirely by hemlock branches, the hemlock's height allows direct light only for a brief period when the sun is immediately overhead. The evergreen nature of hemlocks also provides insulation in the winter months when deciduous trees are dormant. In this way, hemlocks keep streams warmer in winter and cooler in summer, restricting daily temperature fluctuation. The loss of temperature regulation is perhaps the most immediate effect of hemlock loss on streams. Trout are utterly dependent on cold water and the associated high oxygen content for their survival.

At least some temperature buffering will recover as forests regrow. This new tree canopy will hopefully keep streams from reaching temperatures that limit trout populations. However, shading is not the only effect that hemlock has on streams. In many headwater streams, the primary energy source is inputs of organic matter, twigs, and leaves from the surrounding forest, not in-stream photosynthesis from algae. Whether hemlocks are replaced by birches (*Betula* spp.) in the north or *Rhododendron maximum* in the southern Appalachians, the chemical and physical characteristics of litter inputs into streams will undoubtedly change. When leaves and twigs enter a stream, breakdown occurs through a cascade of invertebrates and microbes that collectively form the food web that terminates in the charismatic trout. Hemlock leaves and twigs, with their chemical feeding deterrents, decompose much more slowly, providing a persistent base to this food web. Birch litter, in marked contrast, breaks down much more quickly, leading to a rapid pulse of resources into the stream that supports a different suite of microbes and invertebrates. Wherever *Rhododendron* becomes more abundant along streams, its shorter stature will have less effect on temperature regimes. *Rhododendron* litter is also very resistant to decomposition and persists in streams more like eastern hemlock litter. Of course, *R. maximum* foliar chemistry is quite different from hemlock and will likely lead to other invertebrates dominating the stream community. By changing both the temperature and available invertebrate food base, effects would also be expected on fish communities.

Trout, the raison d'être for eastern fly fishers, are strongly associated with hemlock-covered streams. Brook trout (*Salvelinus fontinalis*) and brown trout (*Salmo trutta*) are more abundant in hemlock-covered streams than those surrounded by deciduous trees. This relationship is not obligatory by any means, so trout would not face extinction, but their populations will suffer. High water temperatures may become limiting to trout toward the southern end of their range, shifting the distribution northward or perhaps into higher altitudes. However, the various trout species effectively represent charismatic megafauna that may make people enthusiastic about conserving eastern hemlock, at least along streams. It is worth reminding that only brook trout is native to Eastern North America. Other trout, brown trout (European) and rainbow trout (*Oncorhynchus mykiss,* western North America), have been introduced into eastern waterways.

Nonaquatic organisms also utilize hemlock. White-tailed deer (*Odocoileus virginianus*) and porcupines (*Erethizon dorsatum*) use hemlock forests as winter habitat and food, though perhaps these animal species do not form the best conservation arguments. We already have entirely too many deer, and porcupines are simply too prickly to be compelling spokes-mammals. Birds, however, capture the interest of a great many hobbyists and are a beautiful and motile representation of the forests that support them. Like trees, there are characteristic birds of early and late-successional habitats, reflecting the niches to which they are adapted. Early successional birds have increased as people have disturbed forests and created semipermanent openings. The agricultural shift toward more productive midwestern portions of the continent has allowed many former New England fields to revert to forests over the last century. Reforestation in the Northeast has ultimately led to an increase in hemlock forests, a process that is now rapidly reversing as the hemlocks die.

Birds that specialize on hemlock forests will decline with canopy hemlock loss; there is no avoiding that. The local abundance of species such as the Acadian flycatcher (*Empidonax virescens*), the black-throated green warbler (*Dendroica virens*), and the Blackburnian warbler (*D. fusca*) is declining along with the eastern hemlock. Whether the loss of hemlock will ultimately result in bird extinction will be determined by whether hemlock-associated species simply prefer hemlock but will shift to other forest types or whether they are truly dependent upon eastern hemlock. Many hemlock-associated birds are not currently abundant, so their ability to persist without hemlock is a serious concern. Of course, some birds, such as the eastern wood pewee (*Contopus virens*) and the hooded warbler (*Wilsonia citrina*), increase dramatically following hemlock loss, but this will likely be a transient pulse until the forest regrows.

## Biological Control of the Adelgid—Additional Invasions

One of the great potential remedies to invasion is the potential to establish biological control. The classic example of biological control is that of the North American prickly pear cactus (*Opuntia* spp.), introduced into Australia's arid regions in the 1840s. *Opuntia* spread rapidly, its seeds efficiently dispersed by birds and potentially competing vegetation cropped back by livestock feeding. Both factors facilitated the cacti to grow unimpeded by resident plants. *Opuntia* covered an estimated 24 million acres (9.7 million hectares) of Australia at its peak, expanding by 2.4 million acres a year. *Opuntia* species have a specialist herbivore in North America, the *Cactoblastis cactorum* moth, also introduced to Australia in 1926 for biological control. This insect feeds exclusively on *Opuntia*, laying its eggs on cactus pads with the subsequent larvae consuming their way through the soft tissue, killing the pad. After just 7 years, the cactus population was decimated. Biological control does not eliminate a species but instead generates cycles between the target and the control agent. Over the long term, this dynamic will generate classical predator–prey cycles, with the populations of both species maintained at low levels.

Effective biological control depends on an enemy organism that specializes on the target for control. In other words, only the invasive target species is consumed, known as monophagy. Without a specialized feeding interaction, there is a risk of damaging organisms that are not the target for biological control. Monophagy can be tricky to assess. The literature describing monophagy in potential biocontrol agents is rife with phrases like feeding "almost exclusively" and release being "virtually risk-free." Mistakes have been made, such as the weevil *Rhinocyllus conicus* released to control Eurasian thistles (*Carduus* spp.) in North America, which subsequently switched to native thistle species (*Cirsium* spp.), including a threatened species (*Cirsium pitcheri*), further jeopardizing the plant's population stability. Potential biocontrol agents often have opportunistic, rapid life cycles that allow them to capitalize on patchily distributed prey, making control quick. The biological control agent also must move and locate the target species in the landscape. In this way, control can be effective over a broad area without requiring land managers to find every target population into which to release the control organism. Finally, the biological control organism must establish a persistent population to provide control continually, functioning forever if necessary. These are the challenges for controlling most biological invasions, including that of hemlock woolly adelgid.

Control of hemlock woolly adelgid in seed orchards, backyards, and other high-priority areas can be easily accomplished chemically with systemic insecticides applied to the soil or directly injected into the trunk. Systemic insecticides nicely restrict effects to insects feeding on the tree—much more focused than other application methods such as aerial spraying. Insecticides can effectively save small numbers of trees but would be prohibitively costly for the millions of hemlocks in a forest. Similarly, some fungi kill adelgids, but these need to be applied directly to infected trees and have little persistence; this is perhaps better thought of as a biopesticide rather than a proper biological control agent. Better options are needed for forested landscapes.

While large populations of sedentary insects would seem attractive to many native predatory insects, appreciable feeding on hemlock woolly adelgid has not been observed. If feeding does occur, it is certainly insufficient to constrain adelgid populations, or they would not have become problematic. Therefore, there has been a rush to identify potential enemies from other areas where the adelgid occurs. While several biological control insects have been assessed, the most promising fall into three groups. *Sasajiscymnus tsugae* is a Japanese coccinellid beetle (the ladybug family) that feeds on adelgids of hemlock, balsam fir, and pines. They are incredibly fecund, laying up to three hundred eggs, and gorge on adelgids as larvae and adults. Their life cycle timing also lines up nicely with the hemlock woolly adelgid, making it a potentially effective control agent. *Laricobius nigrinus*, a derodontid beetle (the tooth-necked fungus beetles—all the other genera in the family feed on fungi), is native to western North America and preferentially feeds on hemlock woolly adelgids but also consumes other adelgids. These beetles are not nearly as prolific as *S. tsugae*, laying only approximately one hundred eggs. However, they are much closer to being native to Eastern North America than the other insect control agents and perhaps represent a safer choice. Finally, a host of *Scymnus* (also called *Neopullus*) species (another coccinellid beetle) from China also extensively feed on hemlock woolly adelgid. Together, researchers have amassed quite an arsenal against the hemlock's enemies.

But is a diverse enemy portfolio essential for biological control? Remember the classic example of biological control—one predator, one prey—a food chain that generates a simple cycle over time. When we introduce multiple consumers, we create a simple food *web* centered on hemlock woolly adelgid. Competition comes into the picture when more than one predator shares the same food base. When there are plenty of adelgids available, two or more species of predatory insects can thrive side by side. However, when the adelgids start to decline, competition between predators will slow their population growth. Competition between predators can destabilize the predator–prey cycle necessary for continued biological control. Insects with rapid population growth may be less able to persist when food availability decreases and their populations crash. Slower-growing predators, or those that can feed on other prey, may be better able to endure, even if they are less able to control the hemlock woolly adelgid. The predatory response of biocontrol insects may be slow enough to allow forest damage when there is an outbreak of adelgids. The slow growth of eastern hemlock populations may make any level of adelgid damage to the trees a serious challenge to sustaining local hemlock dominance.

Many people are rightly dubious about species introductions, regardless of the intent. Introductions, deliberate and accidental, are what generated the forest problems that we face today. Now those invasions are being addressed with further invasions. Globally, governmental agencies have had a bad track record in making introduction decisions. That said, biological control is clearly the best mechanism to save the eastern hemlock, and I am somewhat optimistic it will work. The nontarget effects to other plant pests appear limited, so unless there is an adelgid rights group, of which I am not aware, these introductions feel safe, although no introduced biological control is fail-safe.

## The Future for Hemlock

At this point, a great many established hemlocks have been lost. That will not change. The hemlock woolly adelgid has now colonized much of eastern hemlock's range, so time is short. If the biological control efforts for the hemlock woolly adelgid become effective and persist in the landscape, then recovery of the hemlock may occur. Though adelgids have killed many canopy trees, seedlings and other suppressed hemlocks may remain in sufficient numbers in some areas to allow regeneration. The level of canopy openness may retard their growth initially, but many should survive. The great tragedy here will be the incredibly long time needed for hemlock to resume its position in the forest canopy and recolonize areas where it has been decimated. There is, however, precedence for massive recovery in eastern hemlock. The pollen record clearly shows that the eastern hemlock populations plummeted approximately 4800 years ago, suggesting an environmental change, pathogen, or herbivore outbreak, perhaps in concert. Therefore, our contemporary forests have recovered from previously decimated populations. This fact makes me hopeful for the future. This hope is more for our grandchildren's grandchildren, but it is hope.

Biotechnology may be the next best hope if biological control efforts cannot sufficiently control hemlock woolly adelgids to allow hemlock regeneration. With such a slow-growing and maturing tree, breeding programs to generate resistant trees will take ridiculously long to produce resistant stock for replanting. Eastern hemlock appears an appropriate situation for genetic transfer of whatever confers resistance in other hemlocks (*T. heterophylla* and *T. mertensiana* from North America, plus additional species from Asia). As with American chestnuts (chapter 2), restoration will depend on generating regionally appropriate genotypes for the range of eastern hemlock. Again, the time required for growing from a seedling to a mature tree makes even this a long-term prospect. Such efforts are currently underway.

There has been some movement toward ex situ conservation as a stopgap process—establishing plantations that will serve as the genetic reservoir for reintroduction efforts. If established inside the native range of hemlock, seed orchards will need to be protected from the adelgid, though chemical means would work well in this situation. Developing hemlock orchards outside their native range is somewhat trickier. Western North America has a similar environmental range but also has a diverse adelgid population. For Carolina hemlock, a much rarer species, some have examined South America for environmental matching. Climatic models suggest parts of central Chile and southern Brazil would be ideally suited for eastern hemlock growth. To me, this analysis also indicates that eastern hemlock could be a wildly invasive species in the target area. I certainly hope this plan is not being pursued as an option. However, hemlock woolly adelgid should be a fabulous biological control agent if hemlock became invasive in South America.

Perhaps the most straightforward process would be to transfer one or both western North American hemlock species into Eastern forests. These tree species have resistance to the woolly adelgid, there is a healthy and continuous source for the plants, and they are effectively native, at least to North America. Intracontinental transfer would represent the least expensive route to recovery, though much money has already been spent developing various biological control agents. Transplantation of western species would still take a long time, but all potential options are dependent on equally unsatisfactorily long time scales. The decision to transfer a species is a large one, a choice that should not be determined lightly. I do not think this is a preferred option, but it is an option that must be a part of the discussion.

My sincerest desire is that eastern hemlock restoration efforts will happen much more quickly than those for American chestnut or American elm. One could argue that hemlock is equally charismatic to American chestnut and could garner public and private support. Eastern hemlock certainly is a species that still resides in our mind, with losses very recent in our memory. Trout anglers are passionate and could also contribute to the effort—Trout Unlimited is a nonprofit organization already involved in stream restoration. We have solid templates for tree breeding and restoration from the American chestnut. From those efforts, we can envision the path ahead.

FRAXINUS AMERICANA.L.

Leaves and fruit of white ash, *Fraxinus americana*.

# 4

# WHITE ASH—*FRAXINUS AMERICANA*

Perhaps nothing is quite so American as mom, apple pie, and baseball. Now my mother arguably makes the best apple pie, stacked high with apples and covered with a wonderful crumb topping, and I am fond of baseball. I have frequented baseball games at every university that I have attended. None of the teams have been particularly strong, but the games always provide a relaxing afternoon where you can get some sun and yell at an umpire. Collegiate baseball games are rarely crowded, and the weather can be unpleasant for most of the season in northern states. The one aspect of the game that I cannot stand is that college baseball, in contrast to the major leagues, uses aluminum bats. Quite simply, I cannot tell how well the ball was hit with an aluminum bat. The soulless "plink" that occurs with a home run is amazingly like the "plink" that occurs with a poorly contacted foul ball that goes screaming into the stands. Wood bats provide a much more satisfying crack that honestly represents the contact with the ball. You can hear the difference whether you are close to the infield or in the upper deck of Yankee Stadium, Wrigley Field, or the

Great American Ball Park. The only thing unknown following the sound is whether the ball will make it out of the park or just disappointingly deep into the outfield.

The sound of a hit baseball is a childhood sound, with many households proudly owning at least one Louisville Slugger. Baseball bats, from your first through the ones used by the batters in the major leagues, were traditionally made of straight-grained ash. Ash wood is coarse grained but exceedingly tough and shock resistant. This strength makes the wood ideal for axe and other tool handles, boat oars, and, of course, baseball bats. Sadly, baseball has recently moved away from the traditional ash bat, perhaps permanently now, with the advent of the emerald ash borer (*Agrilus planipennis*) and quarantines that prevent the movement of ash lumber. The current wood of choice is sugar maple (*Acer saccharum*). This wood is also hard but not quite so flexible and shock resistant, leading to a spate of broken bats and some injuries. The change in bat wood has been attributed to Barry Bonds because sugar maple (also known as rock maple) was his choice for many home runs. By the time the sugar maple fad is over in the major leagues, there may not be enough ash wood remaining. Sugar maple is the topic of the next chapter, and this second decline may mean that aluminum alloys will come to Major League Baseball yet. Oak (*Quercus*) and hickory (*Carya*) also can be used for bats (Babe Ruth favored hickory), but these are primarily considered too heavy for today's batters. It is unclear what bats will be made of if both maple and ash wood become unavailable. The great American pastime is changing because of a few, decidedly un-American, beetles; who could have imagined such an outcome?

## Ash Species

Unlike the effectively monospecific chestnut and hemlock of Eastern North America, the ash genus (*Fraxinus*) contains multiple species, separated somewhat by their habitat preferences. Four species are common in Eastern North America. Black ash (*F. nigra*) is a tree of swampy areas with standing water for periods of the year. Green ash (*F. pennsylvanica*) is a species more common to river bottomlands and wetter soils, just not tolerant of flooding to the extent that black ash can survive. White ash

(*F. americana*) appears similar to green ash but occurs in better-drained uplands and is common to midwestern successional habitats, particularly those following agriculture. The last of the eastern species, pumpkin ash (*F. profunda*), is a giant found along rivers and is much less common. I have seen a pumpkin ash only once, as a massive tree next to a river observation deck. The tree was loaded with an abundance of seeds that dwarfed other ash trees' seeds, so there was no mistaking it for another species. The other eastern ash species include Carolina ash (*F. caroliniana*) and blue ash (*F. quadrangulata*). Western *Fraxinus* representatives outnumber eastern species two to one, but most are large shrubs to small trees rather than canopy dominants. The one exception is the Oregon ash (*F. latifolia*), which is large enough to represent a wood source. European (*F. excelsior*) and Asian (*F. chinensis* and *F. mandshurica*) ash trees are planted horticulturally and have escaped locally but never appear to become problematic invaders, at least not yet.

Ash trees are dioecious, meaning they have separate sexes, a relative rarity in plants. Trees are commonly self-incompatible, requiring another genetic individual for successful mating, but having distinct sexes is much less common. Having separate sexes means that successful colonization of a new habitat will require two individuals of the opposite sex, a much more unlikely event. Ash trees become a little more complicated when you get into the species' details. White ash exists as three separate genetic lineages—a diploid (46 chromosomes), a tetraploid (92 chromosomes), and a hexaploid (138 chromosomes). Diploids occur everywhere but are the only lineage at the northern end of the species' range. Tetraploids occur in the southern United States, and hexaploids are most abundant in between. Likely, the species started diploid and had two separate events that generated the other genetic lineages. While they are all considered the same species, they are functionally isolated, as plants with different numbers of chromosomes cannot reproduce with each other. If you stick to the biological species concept, based on the potential for interbreeding, these lineages represent three separate species. Operationally, it would be ridiculous to treat these lineages as individual species, so no one does that. It would take a microscope to determine a tree's chromosome count, which would be impractical in a field setting. Polyploidy is a common way plants produce new species, generating immediate reproductive isolation that allows the chromosome lineages to diverge if

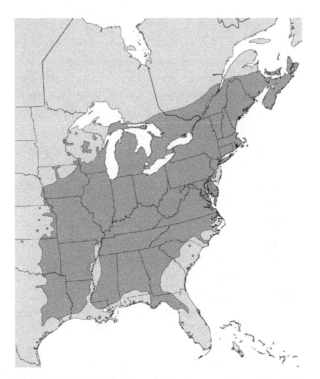

Natural distribution of white ash, *Fraxinus americana*. From U.S.
Geological Survey Geosciences and Environmental Change Science Center.

natural selection pushes them in different directions. This has not happened in white ash but appears responsible for some of the species' morphological variation.

Hybrids between plants with different chromosome numbers have one
mechanism that can make them fertile again, doubling their chromosome
number, another flavor of polyploidy. Regardless of the chromosome
number in each parent, doubling makes the production of reproductive
cells with half the genetic complement of each parent possible again.
Hybridization followed by polyploidization has generated much of the
diversity of plants that we see on the planet. This process may explain
the ancestry of the pumpkin ash. It appears that a tetraploid white ash
crossed with a diploid green ash and doubled its chromosome count
to make a reproductively viable hexaploid pumpkin ash. The process

responsible for producing pumpkin ash is improbable but not impossible. How many thousands of seeds do trees produce every year? Even improbable events become imminently probable over the long run with such numbers at play.

All Eastern North American ash trees can grow rapidly and colonize new habitats. They are all susceptible to the emerald ash borer as well. The remainder of this chapter will primarily focus on the white ash. This species has a broad geographic range and is the most abundant of the ashes. This species, when young, is more shade tolerant than the other ashes, allowing it to regenerate in closed canopied forests more than the others. All ashes can function as classic early successional trees, so similar arguments can be made for each species. Additional details will be added for the other species when they differ substantially from white ash.

## The Role of Ash in Forests

Forests are complex structures containing a tree canopy of various densities with occasional gaps. Gaps are places where a tree once grew, providing canopy openings for light and root gaps for soil resources. Gaps are low competition opportunities that allow trees, shrubs, lianas, and herbaceous plants a window of opportunity before the canopy and roots respond to fill the opening. Gaps vary dramatically in size, ranging from the tiniest opening where a single tree branch fell, to where a single tree died, to where many trees were toppled by a thunderstorm, tornado, or hurricane. Regeneration will vary dramatically with the size of the gap. Surrounding trees may fill small gaps as they grow preferentially toward available light. Gaps of a single, small tree may allow saplings of shade-tolerant trees established in the forest understory to put on a burst of growth until the gap closes. More significant gaps will contain seedlings of fast-growing trees like ash that can capitalize on the abundant resources left behind.

The intermediate shade tolerance of white ash allows it to temporarily persist in a shady forest understory, enabling regeneration in forests that are otherwise dense and closed. Having to colonize a gap from seed may take too long to ensure success. Once a canopy opening occurs, suppressed ash seedlings grow rapidly, outpacing the growth of more shade-tolerant tree species. In forests undisturbed by human activity, such opportunities would be less common. Therefore, the abundance of ash would typically not have been great in pre-European forests. Shade-tolerant species such as sugar maple (chapter 5) respond to gaps in vastly different ways.

With establishment opportunities so rare in the forest landscape, seed dispersal must reach appropriate places for regeneration. Seed dispersal in *Fraxinus* is by wind, with the wings of the samara fruit streamlined and torpedo shaped. When you drop ash seeds, they do not spin or have any other mechanism to slow their descent and encourage seed dispersal. Instead, ash seeds are perfect for dispersal in the higher winds that characterize the borders of large gaps. Their shape aligns them with wind, and the seeds' connections to the maternal plant are broken only by stronger winds that may carry them long distances. Ash trees close to large gaps or blowdown events may produce large numbers of seedlings, which may become canopy trees that persist for a century or more. Trees produce cohort after cohort of seeds to create the next unlikely generation of trees. Unlike many successional trees, *Fraxinus* does not form a persistent reserve of seeds in the soil that functions as an opportunistic reservoir waiting for a forest opening to occur. Ash seeds typically survive only a few years. For the white ash to persist in continuous forests, broad seed dispersal and consistent production are necessary.

Other *Fraxinus* species similarly require openings for regeneration but have additional physiological requirements. For green ash along rivers, flooding events are critical disturbances, and recolonization comes from surrounding trees that escape damage. As flooding disturbances are more common along waterways, green ash abundance would be proportionately larger, mixed with cottonwood (*Populus deltoides*), box elder (*Acer negundo*), and, historically, American elm (*Ulmus americana*). Black ash, characteristic of swampy lands with persistent and frequent flooding, achieves much greater local abundance and is often the dominant tree. The specific habitat requirements restrict black ash to limited occurrence within most forests.

With a specific regeneration niche requiring a large gap, the abundance of white ash was often low in forests. Lucy Braun's work provides an excellent overview of white ash but much less coverage on the other ash species. In her monograph of Eastern forests, *Fraxinus* species appear in all but seventeen of the ninety-one tables (81%) that described forest stands. The majority of the tables that lack ash were in the northern hardwoods forest community, where birches (*Betula* spp.) take over the role of early successional gap filler, or in dry forests, where *Fraxinus* is rare. If we focus on *F. americana*, Braun documented nearly 150 individual forest stands that contained the species. The average abundance of white ash was a paltry 3.7% of canopy stems—not a dominant species by any means. Furthermore, in only eight stands did *F. americana* constitute at least 10% of forest stems, reaching a maximum of 20.9% in one forest. Based on such low abundances in the undisturbed forests that Braun explored, perhaps we should not be concerned about the loss of white ash.

However, farmers have been clearing forests and converting land to agriculture for centuries in North America. Agricultural disruptions to upland forests are essentially anthropogenic analogs for larger gaps in primeval forests that allowed *Fraxinus* regeneration. As such, green and white ash trees have won the ecological jackpot, taking a prominent position as an early successional tree for much of Eastern North America. White ash occupies this successional position alongside trees such as tulip poplar (*Liriodendron tulipifera*), black cherry (*Prunus serotina*), eastern red cedar (*Juniperus virginiana*), and flowering dogwood (*Cornus florida*). Only tulip poplar and ash, however, persist in the forest canopy for long periods. Pines (*Pinus* spp.) are more common in the Southeast, where *Fraxinus* is less common, particularly on nutrient-poor soils. Early successional trees quickly become established after logging or plowing stops, their seeds dispersed by wind or birds. Often the young trees are dense and rapidly form a solid canopy of small trees. Over decades, smaller stems slowly die, allowing a few trees to grow larger, known as self-thinning in forestry and ecological terminology. As ash trees will not regenerate well in a forest understory, more shade-tolerant tree species ultimately displace ash, but perhaps not for a century or more. By then, a new gap or two should have formed, allowing ash establishment to occur. These are not the type of forests that the Brauns visited.

Where I grew up in southwestern Ohio, *F. americana* was by far the most common early successional tree. As people moved out into former farmland from urban and suburban Cincinnati, the trees in their yards were typically white ash, sometimes forming thickets to screen out the road and neighbors. Following road construction, ash colonizes the newly disturbed land, lining highways with their purple and yellow fall color. With the invasion of the emerald ash borer, these trees are dying rapidly now. Problematically, the rapid growth rate of ash trees means that even moderately aged trees can be large and require major equipment for removal. The large size of many ash trees results in much of the financial burden of the emerald ash borer, ultimately paid for by homeowners and municipalities.

## The Emerald Ash Borer—the Beautiful, Winged Harbinger of Death

In the small city of Canton, Michigan, just 8 miles (13 km) west of Detroit, a small green beetle was discovered on some ash trees in the spring of 2002. A little later, the same beetles appeared in nearby Ontario, Canada. Quickly, the ability of this new insect to kill trees became evident. By 2003, populations were expanding across Michigan, northern Ohio, and Ontario. Disjunct invasions also appeared in Maryland and Virginia, likely moved with nursery stock, firewood, or ash lumber. Given the speed at which this insect has since spread, it is safe to assume that the initial infestation was found quickly, within 10 years, likely much less.

The emerald ash borer (*Agrilus planipennis*), is a pretty beetle about half an inch long. It is a buprestid beetle, a family known as the metallic wood-boring beetles or, more poetically, the jewel beetles. Beetles of this large insect family are iridescently metallic. The emerald ash borer has brilliant metallic green outer wings with a bronzy-gold iridescence. In northeastern China, Japan, and surrounding areas, the beetle's native range, the beetles feed on *Fraxinus chinensis*, *F. mandshurica*, *F. rhynchophylla* (sometimes listed as a subspecies of *F. chinensis*), and perhaps other ash species. These ash species appear to have evolved some resistance to the beetle, likely a chemical defense. Under most circumstances, infestations in the native range are light, with the beetles preferentially attacking stressed trees in a population. However, outbreaks have occurred in the native range, suggesting the potential for the rapid and devastating spread seen in the New World. It is possible that an initial outbreak occurred on North American ash trees planted in China that were salvaged for lumber. Using this lumber for pallets or other packaging may have paved the way for the borer to enter shipping routes via infected wood.

The life cycle of the emerald ash borer is typical of the entire buprestid beetle family. In June or July, the beetles lay eggs in cracks of the bark of susceptible host trees. A week or so later, the eggs hatch and the larvae chew their way into the nutritious phloem and cambium tissues beneath the bark. For the remainder of the summer, the larvae chew through these tissues winding back and forth, the tunnel expanding as the larvae grow and molt through four developmental stages. The snake-like gallery fills behind the feeding larvae with frass, the technical term for bug poop, potentially leading to secondary infections. In the late summer or autumn, mature larvae chew out a pupation chamber under the bark, where they will overwinter. Late the following spring, the larvae will pupate, metamorphose into adults, and emerge from characteristic D-shaped exit holes that they chew in the bark. Adults then feed on ash foliage, mate, and start the cycle all over again. Each female may lay up to two hundred eggs, producing the potential for massive population growth. The life cycle may take 2 years to complete in northern populations because of the shorter growing and feeding season.

Though adult emerald ash borers feed on *Fraxinus* foliage, that damage is trivial. The critical damage occurs because of cambium and phloem destruction. Over time, this damage girdles more and more of the tree, generating water stress and leading to canopy thinning as branches die. Bundles of stem sprouts below girdled locations, like those of chestnut blight, also commonly occur. Initial borer infections often occur in the upper canopy, with subsequent generations moving down the tree. There may be little visible evidence in the first year of illness. By the time characteristic D-shaped exit holes appear at eye level, tree death is likely imminent. Trees typically die within 4 years, and as little as a single year when massive numbers of borers attack a single tree. When an ash tree dies, it may possess between fifty and eight hundred emergence holes, each representing a reproductive adult beetle in the landscape.

As with all beetles, adults fly to seek out mates and to find host trees. While most movements are local, flights of half a mile or more are possible within a single day. Most beetles fly only a few hundred feet to find the next host tree. With such a small beetle, movement is also facilitated by wind, particularly from approaching weather fronts and storm events. Spread from an initial infestation has been estimated at over 6 miles (10 km) per year, so longer movements are common. Long-distance movement often occurs with the transportation of infected ash wood, the likely mechanism of the initial transfer from Asia. When a tree dies in a yard, it commonly ends up as firewood. Families frequently take firewood on camping trips, often to locations surrounded by trees where the adult beetles may emerge and colonize. Wood movements enhanced the early spread of the emerald ash borer and were likely necessary within the agriculturally fragmented forests of the Midwest.

The spread of the emerald ash borer epidemic is disheartening in every aspect. Though stressed trees are typically more susceptible, they are often just the first trees attacked. All New World *Fraxinus* appear to be vulnerable. In the northern forests around Michigan that were first colonized by the emerald ash borer, the three most common ash species co-occur— *F. americana*, *F. nigra*, and *F. pennsylvanica*, allowing comparison of their susceptibility. Black ash (*F. nigra*) succumbed first, followed by the remaining two ash species. Disease ecology teaches us that high host density or abundance should increase transmission rates; however, one ash study found no density effects and another found greater insect damage in low-density ash populations. Community ecology teaches us that forest diversity should offer protection because the hosts become more challenging to find—analyses found no support for this effect either. Suppressed individuals beneath the forest canopy are sometimes the first to succumb to the emerald ash borer, and sometimes they are not. Quite simply, emerald ash borer colonizes a forest stand, attacks *Fraxinus* trees from large canopy individuals down to individuals only an inch or so in diameter, and kills nearly all of them within 6 years—often less. Variation in susceptibility among habitats and species is purely academic, as almost all *Fraxinus* trees die.

There is some indication that emerald ash borer will be sensitive to the cold, potentially saving more northern populations of *Fraxinus*. With climate change making low winter temperatures less predictable, this hope is not entirely reassuring. Furthermore, sexual reproduction in northern populations may allow selection for enhanced cold tolerance. A few warmer than average winters, along with evolving a bit of additional cold tolerance, may be all the borer needs to reach the northern extent of *Fraxinus*. Nothing but the distribution of ash trees appears to limit emerald ash borer's spread to the south. As of 2022, emerald ash borer has made the jump to western forests and is now found in Oregon, threatening western species of ash.

## Forest Recovery following Ash Decline

A critical aspect of emerald ash borer invasion is whether ash will persist in any of the forests following an outbreak. Most studies have documented a canopy ash tree or two that survive with relatively good canopy health late into a local borer invasion. Whether these trees represent individuals genetically resistant to the borer, in which early borer damage induced resistance, or individuals that just take longer to die is not yet clear. If there is some resistance in native ash populations, breeding programs could exploit it to produce resistant trees. More importantly, natural interbreeding of two ash trees with resistance could result in a flush of resistant seedlings dispersed into the newly opened forest landscape—the perfect environment to enable a cohort of ash to become established. As *Fraxinus* grow much more quickly than eastern hemlock (chapter 3), recovery could occur on a time scale of decades rather than centuries. Most people do not seem to hold out hope for this scenario, but I think it is always good to hope.

A slightly more plausible source of *Fraxinus* regeneration lies with the fate of the remaining ash trees that are too small to support an ash borer when the outbreak sweeps through a forest stand. Canopy trees are the source of seeds to produce new seedlings. As canopy trees succumb to the emerald ash borer, seed production also decreases. Ash seeds do not persist for more than a few years in forest soils, so germination and mortality will rapidly deplete these seeds within a few years of canopy death. As most researchers assume no canopy ash trees will survive, there will be a narrow chronological window for new seedlings to grow into small saplings under an open canopy.

The fate of the post-borer ash generation will ultimately depend on the population of the emerald ash borer following the loss of its primary hosts. Emerald ash borers persist in low densities following the loss of canopy trees, but populations are often evaluated immediately following the death of most ashes. How long low insect densities will be able to persist is unknown. Each year, a few small saplings will achieve sufficient size to be attacked by emerald ash borer, but each of these cannot support more than a few borers through to maturity. As time passes, limited borer reproduction may lead to localized extinction, particularly as the wave of outbreak tends to attack all available hosts in a region. Therefore, recolonization of individual forests by the borer may be sufficiently rare to allow ash regeneration to occur. Again, most people do not appear to hold out hope for this scenario, but we simply do not know how well emerald ash borers will persist long term under low host tree densities.

The impact of *Fraxinus* loss will vary with the system involved. In mature upland forests where *F. americana* is only 3% or so of the forest canopy, ash loss will have minor effects, with the gaps left behind filled in by any of the several species that co-occur with ash trees. In early successional forests where ash species may be abundant, large canopy openings will develop. If an understory of shade-tolerant tree seedlings had already developed, the loss of early successional tree cover might accelerate succession toward a more mature forest composition. In forests that are still young, forest regrowth will effectively be reset, allowing early successional trees such as black cherry, elm (*Ulmus* spp.), and eastern red cedar to colonize. For the ash trees of wetter environments, where the species also tend to dominate, larger openings will likely occur. This canopy opening would result in a pulse of recruitment of fast-growing wetland trees—cottonwood, box elder, and American elm (where it persists), with the addition of red gum (*Liquidambar styraciflua*) farther south.

Though forest biomass and stature will certainly not achieve pre-borer levels for decades, canopy closure should happen relatively quickly. With late-successional replacements available throughout ash's range, forest regeneration should still proceed as usual. The one place where I see potentially persistent impacts would be in riparian and wetland forests where *F. nigra* and *F. pennsylvanica* dominate. *Fraxinus* species are considered important for wildlife because of the seeds they produce, but that is similar for almost any tree that produces seeds a small mammal can consume. However, the species likely to replace ash in wetter habitats are not equivalent food sources, producing either tiny seeds (cottonwood, red gum, or elms) or producing them in the spring (boxelder), when small mammal populations are low. Unfortunately, in *F. nigra* swamps, the openings left by the loss of black ash have not filled in naturally and may require active mitigation—the planting of appropriate trees.

Last, I must note that the contemporary abundance of *Fraxinus* in the modern landscape of North America is essentially a result of human impacts on that landscape. Successional habitats, and therefore ash habitats, are primarily generated through human activity such as logging, agriculture, or road building. These disturbances and the propensity of ash to disperse large distances have generated massive numbers of *Fraxinus* in North America—estimated to be over eight billion trees and saplings in the U.S. Forest Service's Forest Inventory and Analysis, the authoritative national record for extant forest tree cover. The forest recovery suggested above depends on trees being able to regenerate. When young forests with large populations of *Fraxinus* are also invaded by non-native species, the exotics may gain the upper hand. Exotic shrubs such as bush honeysuckle (*Lonicera* spp., chapter 7) can expand rapidly in the open canopy left following borer outbreaks. These shrubs may compete strongly with tree seedlings, slowing or preventing forest regeneration.

Ash trees have been planted widely in cities, mine restorations, and riparian corridors. The loss of canopy American elms generated many opportunities for ash trees; the loss of chestnut offered the same prospect decades earlier. Linkages among tree species are essentially unknown features of our forests—change one tree, and the resulting long-term impact is as yet unidentified. The additional open-ended question posed by non-native plants may change things further in ways that are still unclear. Alter as many forest species as we have over the past century, and anything may happen.

## Quarantines—the First Step in Control

Once we realized the risks, the initial response to emerald ash borer was a quarantine to manage the spread of the insect, slowing it until management options could be developed. High-intensity effort extirpated a few outlier populations, but the effort needed was impractical for the heart of ash's range. Wood products or nursery stock of ash trees could not be shipped from within the quarantine to areas outside the quarantine. Quarantined locations were based on wherever the borer was found or areas likely to be invaded next. Rapid work ahead of the quarantine provided landowners an opportunity to harvest any trees for profit while they still could. The quarantine also resulted in a "don't move firewood" campaign to educate the public of the unintended consequences of such a simple action.

One of my former students took a position with the U.S. Department of Agriculture as a plant protection officer. This job was much more that of a law officer than I originally expected as it involved months of extensive training, including firearms (people get touchy when you take away their work—thus the need for a gun). His job was to contact plant nurseries and wood product manufacturers to ensure they were aware of and complied with the quarantine. While nurseries are easy enough to track, many independently run lumber outfits have mobile sawmills that make pallets for shipping. These producers were much more challenging to track down and even trickier to work with as there is no way to tell where a pallet can end up. As an agent, he caught a quarantine violation that regulations required returning the offending plant material to Illinois. Ironically, while moving plants to an area outside the quarantine might expose all forests along the way to emerald ash borer, putting the plants and any accompanying beetles back on a truck to return to where they were produced would do the same again. The approach did not appear to follow common sense.

Managing a quarantine depends on good location data for the emerald ash borer. Monitoring of the emerald ash borer outbreak is primarily through finding infected trees and setting purple box traps for the adult beetles. The traps are a triangular column of purple corrugated plastic, a color preferred by female emerald ash borers, and baited with attractants. Stressed ash trees produce chemical signals that leak out and diffuse into the atmosphere, attracting the emerald ash borer. As the *Fraxinus* species in the borer's native range have some defense against the insect, selecting stressed trees means choosing a host less likely to protect itself. North American *Fraxinus* species have little resistance to the emerald ash borer, but the insects' chemical cues are evolutionarily ingrained. These chemical cues are exploited to lure the insect into monitoring traps. These traps were set out in astonishing numbers—22,000 in 2014 alone—concentrating trapping efforts in areas adjacent to known populations. For years, you might have seen these traps dotted along local highways in the eastern United States, and then they were all gone in an instant. Once the insect is present in an area, the critical zone for sampling moves to where the insect has not yet been detected, and the traps must be moved to catch the borers before they break out in a new region.

## Biological Control—Building Biotic Resistance

Some biological control of emerald ash borer naturally occurs as wood-peckers and other bark-foraging birds feed on borer larvae before the adults emerge in the spring. Birds may consume 45% or more of emerald ash borer larvae on trees with thinning canopies. More important as a natural control, birds focus their foraging on trees with the highest levels of canopy decline, an indication of greater infestation levels. While the in-crease of woodpecker food resources is likely not to last, the bird's feeding activity can reduce insect spread. Land managers can leave standing dead trees (snags) to encourage woodpecker nesting and foraging to help miti-gate emerald ash borer invasion. Leaving snags is always a good idea from a forest management perspective, but this would need to occur before the borer arrives to increase the bark-foraging bird population. Ash trees typi-cally fall within 4 years of dying, so snags left by the borers are unlikely to last as nest resources for woodpeckers. There is evidence that bark-feeding birds are increasing with the additional winter foraging resources, when woodpeckers are more dependent on insect larvae.

More traditional biological control has focused on introducing par-asitoids that feed on emerald ash borer in its native range. A parasite, as most know, is something that completes its life cycle living inside a host, like a tapeworm. Parasitoids differ in that parasitoids lay an egg on the host insect, and the parasitoid larva consumes the host as they both develop. Parasitoids are common in integrated pest management as they can be host specific and are excellent at finding and targeting their hosts. An additional challenge for would-be parasitoids of emerald ash borer is that ash borers develop under a thick layer of tree bark. Fortunately, parasitoids on bark-dwelling insects are often quite adept at using chemical cues to locate their hidden hosts. They also have needle-like ovipositors long enough to penetrate bark to get an egg on the beetle larvae beneath.

As it turns out, the emerald ash borer has an impressive array of both native and introduced hymenopteran (bee and wasp) parasitoids. Emerald ash borer supports more than twenty parasitoid species in seven insect families. The parasitoids run the full spectrum of life histories of parasitoids. There are ectoparasitoids that dwell on the host's external surface and endoparasitoids that live inside the host. Some are solitary, a single parasitoid per host insect; some are gregarious with multiple parasitoids per host. There are egg parasitoids, and there are larval parasitoids. Native parasitoid insects are generalist on bark beetles, using multiple species as host, so their life history does not precisely align with the emerald ash borer. Most native insects parasitize only a small percentage of the available emerald ash borer larvae, so their utility in achieving population control is doubtful. If the native parasitoids could control emerald ash borer, there would never have been a population outbreak since the parasitoids would have limited the invasiveness of *A. planipennis*. The beetle would have remained just another inconsequential, non-native insect.

The immediate response to the invasion of emerald ash borer was a rapid search of Asian ash populations, which yielded three potential biocontrol insects. These were reared, approved, and released in 2007, just 5 years after the initial identification of the borer outbreak, a remarkably fast response. All the intentionally introduced parasitoid insects are specialists, or at least appear to be, on emerald ash borer. *Oobius agrili* (Encyrtidae, a chalcidoid wasp), is an egg parasitoid. Egg parasitoids appear ideal to control the insect as their effects occur before any real damage to the ash tree. Adding to its convenience, *O. agrili* is parthenogenetic, reproducing asexually the way that the hemlock woolly adelgid does. This characteristic should make *O. agrili* able to increase rapidly when emerald ash borer densities are high.

While emerald ash borer eggs are easily accessible, accessing the larvae requires getting into the host tree's bark. *Tetrastichus planipennisi* (Eulophidae, another chalcidoid wasp) is a tiny parasitoid, around an eighth of an inch long. This parasitoid lays several eggs on each emerald ash borer larvae. Unfortunately, the insect has a short ovipositor, limiting its effectiveness to small ash trees with thin bark. *Spathius agrili* (Braconidae, an ichneumonoid wasp), the last of the species released in 2007, feeds on older emerald ash borer larvae, producing five to six adult *S. agrili* per larvae killed. This insect can better deal with emerald ash borers in ash trees with thicker bark but appears cold limited and thus less effective in the northern region of ash's range. A closely related insect, *S. galinae,* appears to be hardier for northern climates and was approved in 2015 and first released in 2016.

Just as an effective quarantine depends on excellent monitoring data, so does evaluating biological control introductions. The problem arises with the tiny size of most parasitoids and their location—on or under the bark. Flying adults can be lured to traps, but this is a general approach for many insects, yielding scads of dead insects to sort through. You can also place out "trap" logs with *A. planipennis* eggs or larvae in them, but this is very labor intensive. Because of this, detailed work has focused on more direct and destructive measurements that often involve chainsaws.

To look for the egg parasite *O. agrili*, you must examine each borer egg individually—they conveniently turn different colors when parasitized. You can either collect eggs in the field or strip bark off trees with a drawknife, bring it back to the lab, and isolate the eggs. Regardless of the technique employed, it is a lot of work. Assessing larval parasitoids is always destructive. Researchers must cut down ash trees, peel the bark, and examine each emerald ash borer tunnel to determine whether it had produced an adult borer, had a living but uninfected individual, or was parasitized. This work is necessary to determine whether the introductions have been successful, whether the parasitoids can pursue emerald ash borers throughout the landscape, and which biocontrol insects are responsible for mortality. If this process seems like it would take a lot of time, energy, and money, you are correct. It is the only way, however, to assess success and to adapt management strategies regionally.

Has this work paid off? It certainly appears so. The parasitoids have become established, except for *S. agrili* in northern areas. More importantly, the parasitoids have dispersed throughout the landscape and have appeared places where parasitoids were not released. From the perspective of the emerald ash borer, these introductions have decreased their population growth over time. Control could happen and maybe even is occurring now. Subordinate ash trees have survived in some forests and appear to benefit from biocontrol efforts. At the margins of the borer invasion, spread seems to have slowed. When the borer appeared locally in east-central Illinois a few years ago, I expected rapid decline and death of trees. Mortality did not happen as quickly as suggested by the pre-biocontrol data, though. Based on these observations and the data available, I cannot say that biocontrol has been wholly successful as trees are still dying, but control may be on the horizon. Parasitoid insects continue to be reared and released yearly, with more potential insects for release explored. Eventually, the invasion front of the emerald ash borer may be mirrored by the invasion front of the controlling insects, preventing borer infections on all but the most stressed trees. When or if that happens, successful biological control of the emerald ash borer will have been achieved.

## The Costs

I have seen claims that the emerald ash borer invasion is the most expensive insect invasion yet. While I understand the need to quantify impacts, both financial and ecological, I firmly reject the desire to qualify one invasion as more or less expensive or impactful. Extinction, even functional extinction, is forever. Any annual financial gain associated with a tree species, amortized forever, essentially results in infinite income over the span of extinction. Therefore, economists apply discounts to future values, estimate costs, and draw upon additional financial manipulations. Movie accountants similarly try to show the value of a feature film, by making such claims as, "It had the highest-grossing ticket sales ever for the third quarter of the year." However, all things are not equal in such comparisons. Ticket prices have increased, as has the population. Inflation plays havoc with the actual value of the dollar spent, and the cost of making a movie goes up. Therefore, ticket receipts should increase over time. Value-based judgments such as the "best" or "worst" will always be qualitative, regardless of how a position is justified.

Some costs are knowable, some are not. With the decline in *Fraxinus*, some costs can be clearly assessed. Most direct and most immediate are the costs of removal and replacement of trees. As one of the most common tree genera planted in urban areas, many cities and towns have an abundance of ash trees, often susceptible native and European varieties. When emerald ash borer first entered an area, before biological control efforts began, nearly all the ash trees would die within 4 to 6 years. As *Fraxinus* typically fall within a few years of dying, removal in urban areas is an immediate necessity. Removal costs will be reflected in local budgets and can be estimated for homeowners. Trees in cities and towns provide clear ecological and public health benefits, but their economic impacts can be more difficult to calculate. Trees provide shading to reduce cooling costs, reduce air pollution, take up water to control stormwater runoff, increase property values, and have mental health value, tending to make people happier and more content. The critical variation of note here is that municipalities and individual homeowners bear the costs of removing the dead and dying ash trees and their replacement. Most of the benefits of trees, however, are accrued at a neighborhood/community level. The discrepancy between who bears the cost and who reaps the benefit is always a major challenge in facing environmental problems.

The cost of tree replacement is also knowable for municipalities and can be estimated for homeowners. For many cities, the number of ash trees that will die will represent a monumental task for city workers, one that will need to be spread over several years for logistical and financial reasons. Nurseries will need to increase production to meet the increased volume of replacement trees required and will likely constrain the diversity of replanted trees to those most easily propagated. Prioritization of areas for replanting efforts will almost certainly generate discussions of social justice. The health and ecological benefits of the replanted trees will take many years to recover to pre-ash borer levels, regardless of whether we ever understand the ultimate costs.

Additional costs of the loss of ash will be in the value of forest products lost—the wood. In the short term, there will be a rush to preemptively harvest trees before emerald ash borer arrives or salvage timber after the insect has colonized. This activity should lead to an abundance of ash wood, reducing its value briefly, as seen with the chestnut blight (chapter 2). Except for its use as tool handles, and baseball bats, there is little specialized usage of ash lumber. Surplus wood supply will almost certainly mean an abundance of firewood and pallets if loggers can avoid issues with the quarantine.

Emerald ash borer also attacks one additional tree species, *Chionanthus virginicus*, the white fringetree. This pretty native of the southeastern United States is a small understory tree or a large shrub planted as an ornamental north of its native range. The name *Chionanthus* comes from the Greek *chion*, meaning snow, and *anthos*, for flower, providing a good idea of what they look like in the spring. While a threat to a native plant species is always concerning, more so is the ability of emerald ash borer to complete its life cycle on another member of the Oleaceae (the olive family). Ash represents the primary tree species of the family, but the family also includes many common shrubs, including privet (*Ligustrum*, invasive), lilac (*Syringa*), *Forsythia*, and jasmine (*Jasminum*), and, of course, olive trees. Emerald ash borer has already contacted at least some of these species, so perhaps this worry is unfounded. Lilac would seem to be the one eastern plant that produces stems sufficiently large to support borer larvae, and people would likely notice if their lilac border suddenly died. Experiments have shown that the borer can reproduce on cultivated olives, potentially representing a second massive economic impact of the insect's invasion.

## The Future of Ash in North America

It is entirely too soon to say with any level of confidence what the future holds for *Fraxinus* trees in North America. Fortunately, the outbreak was detected quickly, and the critical work to address the invasion has swiftly happened in response. While the quarantine failed in containing emerald ash borer, no one ever thought that containment was genuinely possible. Instead, the goal was to slow the inevitable spread to allow other work to occur. It is impossible to know how much more quickly emerald ash borer would have spread without the quarantine in place. Every year delay in the spread of the borer was more time to mount a defense. We now know that sylvicultural treatments such as forest thinning are largely useless in stemming the tide of the insect. Releasing the parasitoid species was key; releasing them in enough locations along the invasion front will be the next critical step. It will be years before we know the ultimate effectiveness of biological control agents not only in slowing the invasion of emerald ash borer but, more importantly, in maintaining emerald ash borer at levels low enough to allow *Fraxinus* regeneration.

Perhaps this outbreak will naturally run its course and die out all on its own. Maybe biocontrol will prevent the spread of the borer throughout the entire range of *Fraxinus*. Possibly nothing will work, dooming ash trees to become oddities in botanical gardens maintained with annual applications of pesticides. Even if the emerald ash borer is stopped tomorrow, massive areas of forest have already been affected. These forests will take a while to recover naturally, and some may require intervention such as assisted migration (transplanting ash trees) to regain forest function. The true impact of the emerald ash borer outbreak will not be known for decades, if ever. This time is necessary for forest regeneration and all the plant and animal species that ash support to respond fully. And, I would like to add, this assumes that we will document what happens long enough to understand and learn from it.

The speed of response to emerald ash borers has been encouraging to me. This work, more than anything contained in this book, has shown me that we can rise to the challenge and address forest pests when sufficiently motivated. Quick action appears to have generated a response that can establish lasting biological control on the non-native borer. However, the question remains as to why this tree threat motivated us to act so quickly while others have not. The next chapter deals with another beetle that has also generated a rapid, scorched-earth-type response. This beetle also likely moved in with wood packaging materials and is a significant threat to the trees of Eastern North America. While I cannot help but be enthusiastic and thrilled with our responsiveness to emerald ash borer, I am equally frustrated by our inability to address the primary mechanism that repeatedly allows these insects to colonize from around the globe.

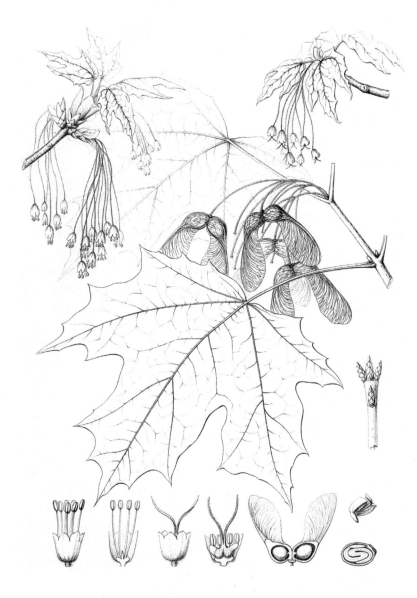

ACER BARBATUM, Michx

Leaves, flowers, and fruit of sugar maple, *Acer saccharum*
(formerly *A. barbatum*).

# 5

# Sugar Maple — *Acer saccharum*

Sugar maple has always been one of my favorite trees. They are extraordinarily shade tolerant, developing layer after layer of thick leafy canopy that casts a dense shade on the forest floor beneath. I grew up climbing sugar maple trees. Unlike less shade-tolerant trees, sugar maples do not self-prune much, retaining lower branches. Suppose that at least the early years of a sugar maple's growth were in open canopy conditions. Many lower limbs in that situation will be large enough to support weight— making a very climbable tree. Shade-grown sugar maples are essentially poles with tiny side branches useless for climbing unless you can shimmy straight up, which I never could. One time I wedged one foot into the topmost fork of my favorite climbing maple, curled my other leg around, and stuck my head out of the top of the canopy, the tree gently swaying beneath me. It was glorious, idiotically dangerous, but a purely magnificent experience.

I was fortunate enough to grow up close to Hueston Woods, an old-growth beech–maple forest that Lucy Braun sampled and included in her book. Beeches remain smooth barked throughout their lives, a lovely contrast to the rougher bark texture of sugar maple. While I did not learn how unique Hueston Woods was until much later, I was immediately smitten with this forest. It was, in the grandest sense of the phrase, a cathedral of trees. The massive beeches and maples dwarfed what I had thought were perfectly reasonable trees in the second-growth forest near my childhood home. In that childhood forest, the largest trees were oaks, with only a couple more than 2 feet (0.6 m) in diameter. In marked contrast, the trees in the old-growth stand were often 4 or more feet in diameter. They were widely spaced, with a shorter next generation of trees in the subcanopy, awaiting their opportunity should one of the giants fall. I fell in love with that magic forest as a child; I grew to appreciate Hueston Woods as I learned what a rare treasure it was. I still visit that southwestern Ohio forest whenever I can. The forest retains its magical hold on me, but I now see the invasive plants that have colonized it and worry about the forest's future.

Sugar maples' beauty is expanded dramatically in the fall when their canopies are adorned in peachy, pale-yellow foliage. They provide much of New England's famed fall display. Maples are also marvelous in the spring. Early, when the sap rises in the trees before the new leaves expand, you get a wonderful treat. Trees accumulate minor damage all winter but have no way of healing themselves as it is too cold for cell growth, leaving many small, broken twigs. When the sap rises in spring, it leaks out of the broken stem tips, producing a slight rainfall of tree sap gently falling on the previous summer's fallen leaves. Even better, the air fills with a subtle maple syrup smell. It is a beautiful yet fleeting forest moment. The sap rise is over quickly, and the trees repair their damaged tissues as temperatures rise and metabolic activity resumes.

Beyond maple syrup, sugar maple wood is economically valuable. The wood is hard and closed grained, making it ideal for furniture and instrument making. It is light colored when the wood is new; over time, the wood will deepen to a golden-yellow hue. Our first home, constructed in 1900, had maple floors that were beautifully honey golden. Furthermore, the builders laid the flooring in concentric bands that started at the perimeter, winding inward until a small rectangle remained in the center of each room. Grain variation in the wood makes some sugar maple trees even more valuable in furniture making. Wavy grain patterns, known as quilted maple, and small circular swirls of grain, known as birds-eye maple, are greatly valued as accent veneers.

Sugar maple, unlike the rest of the trees in this book, is facing a potential threat, not range-wide decimation. This insect threat has been successfully rebuffed, or at least contained in North America. However, if introductions of the pest keep occurring, and this seems likely, the insect may eventually become fully escaped and unstoppable, as happened with the emerald ash borer (chapter 4) and the hemlock woolly adelgid (chapter 3). For this reason, we will spend much more time in this chapter discussing what fabulous trees sugar maples genuinely are. We will return to the pest that threatens this tree and discuss why this insect may be the most devastating yet to the forests of Eastern North America.

## Maple Syrup—Why Not Oak or Ash or Beech?

Every day we, as mammals, tempt ourselves with sweet treats, sometimes succumbing, sometimes not. What is the source of all this sugar? Relatively few plant species produce significant amounts of free sugar as it turns out, at least naturally. In North America, the sources are sugar beets (*Beta vulgaris*, essentially a variety of the regular garden beet), sugar cane (*Saccharum officinarum*, a grass), sorghum (*Sorghum bicolor*, another grass), and sugar maple trees. We can chemically convert any plant-generated starch into sugars, such as industrial corn syrup. Still, few plants produce sugars in sufficient amounts to be a direct sugar source. Why is this? Most sugars are converted into permanent structural carbohydrates, such as cellulose, or condensed and stored as starch or oil. Plants can draw upon stored carbohydrates whenever the plant requires carbon, either for respiration or for growth and conversion into structural components. Most importantly, carbohydrates stored this way will not alter the multitude of cellular processes that occur within the plant.

Accumulations of appreciable amounts of sugars are relatively rare because of the chemical problems that they cause in plant tissues. Dissolved sugars draw water across membranes (osmosis), leading to either increased internal pressures for the cells containing the sugars or potentially catastrophic water withdrawal from cells that do not. The high solute concentrations also interfere with many cellular processes, essentially making metabolism impossible.

Because of this, the production of sugars typically is the last phase in fruit ripening other than the color change, which signals "eat me" to any passing frugivore (fruit eater). We have traditionally used sugar as a preservative for the same reasons—bacterial or fungal cells that land on jam immediately have all their internal water withdrawn and die. Salting for preservation follows the same basic logic. If your favorite fruit preserves claim to contain no preservatives, know that the claim is patently false.

The metabolic problems generated by sugars explain why so few plants produce sugars in high concentrations. Why then is sugar maple unique as the only tree that we exploit for sugar on a large scale? It has to do with the release of sugars during the brief window when the sap rises. Actually, rising sap is an incorrect generalization—sugars are stored as starch in living cells in xylem rays in all active woody tissues, from the tiniest canopy twigs aboveground to minor roots belowground. These rays move material horizontally, perpendicular to the up/down orientation of most of a tree's vasculature. Starch stored in the cells is converted back to individual sugars when temperatures are below freezing at night and slightly above freezing during the day. These sugars pull a lot of water into the tissues, generating pressure in the xylem. If you "tap" a tree, you drill a small hole into the outer layers of xylem, the sapwood, where the sugar-laden sap is pressurized. Attach a spigot, and you are collecting very dilute maple syrup with a sugar concentration of 1–5%. Collect around 40 gallons of sap to make a gallon of maple syrup. I have successfully done this with the one sizeable sugar maple on our land—it is delightful!

Maples (*Acer* spp.), along with walnuts and butternuts (*Juglans* spp.) and birches (*Betula* spp.), are among the only trees that load their xylem with sugars, pressurizing them early in the spring. Species other than sugar maple can be used to make syrup, but the sugar concentration is much lower, so using other species is uncommon. Xylem loading may offer cold protection for trees growing in northern climates. Sugar depresses the freezing point of sap, providing some protection from freezing and its damage. The greater the sugar concentration, the lower the freezing point goes. If freeze protection were the only important factor, we might expect this phenomenon to be much more common in northern forests. Instead, I think a better argument can be made for healing the vascular tissue in preparation for the spring flush of growth.

During the winter, water-conducting xylem cells accumulate many embolisms—gas bubbles in the vessels that interfere with their function. Hydraulic systems work because liquids cannot be compressed; straws work because you can put liquids under tension. If an air bubble forms in a xylem cell, the cell becomes nonfunctional because the air bubble expands as water is removed, preventing additional water from being pulled up to replace the transpired water. In xylem cells pressurized in the spring by releasing sugars and the accompanying water, gas pockets become compressed and can be re-absorbed, healing the embolism and restoring function. This process allows the plant to repair accumulated winter damage and rapidly mobilize resources to produce the new season's leaves and flowers. It is an ingenious cold-tolerance strategy.

## Sugar Maple in Our Forests

Over the centuries, sugar maple (*Acer saccharum*) has been called by various names, not surprising for a tree with such an expansive range. Typically, black maple (formerly *A. nigrum*) is considered a subspecies of sugar maple (*A. saccharum* ssp. *nigrum*); historically, the reverse was also true. This taxonomic merger is just as well, as I could never reliably separate black and sugar maple. The discriminatory trait is whether the last two lobes of the leaf are bigger (sugar) rather than smaller (black)—way too subjective to be helpful for me.

With the rush to explore the New World botanically, the publication of floras was too slow to keep up, and scientists applied a diversity of other names to the species. *Acer saccharum* is considered the correct name because it was published first. In 1785, American botanist and cousin to the Bartrams (see The Changing Context of North American Forests in Introduction), Humphrey Marshall officially named the species. Many consider him the originator of American dendrology. In contrast, *Acer barbatum* (used in Sargent's work, which illustrates this chapter) was coined in 1803, when French botanist André Michaux posthumously published his *Flora Boreali-Americana*. Other names applied to sugar maple and its varieties over the years have included *A. palmifolium*, *A. saccharophorum*, *A. subglaucum*, and *A. treleaseanum*.

Natural distribution of sugar maple, *Acer saccharum*. From U.S.
Geological Survey Geosciences and Environmental Change Science Center.

Mature sugar maple trees in old forests are typically 300–400 years
old, 90–120 feet (27–37 m) tall, and 3 or more feet in diameter. Giants
can live significantly longer and reach almost 7 feet (2 m) in diameter.
With such an extensive geographical range, sugar maple is an important
component of many forests—sometimes achieving dominance, sometimes
not. Lucy Braun's description of deciduous forests reveals sugar maple's
ubiquity in Eastern forests. Sugar maple is one of the many species of
the diverse mixed mesophytic forests of the southern and central Appalachians. In more mesic forests of that region, it is often one of the most
abundant species in the canopy. As you move westward from mixed mesophytic forests, the forests transition into beech–maple, bounded by the
limit of the Wisconsin glaciation on the south, and oak–hickory forests
to the west, as the climate becomes drier and historical fire frequency

increases. Though codominant in beech–maple forests such as Hueston Woods, American beech (*Fagus grandifolia*) is typically more abundant in the canopy, often outnumbering maple 2:1. Within the more western oak–hickory forests, sugar maple was present historically at low abundance, but more about that later. To the north, maple–basswood (*Tilia* spp.) forests occur in a narrow band primarily in Wisconsin and southern Minnesota. Sugar maple is also an important component of the northern hardwoods. Throughout its range, sugar maple is common in forest understories, yielding canopy position to super-shade-tolerant species such as beech and hemlock (*Tsuga canadensis*) wherever they dominate. All maples are intolerant of fire, which is probably their more critical restriction on dry sites than actual water stress.

## Sugar Maple Regeneration—Made in the Shade

The shade tolerance of sugar maple places it in close similarity with two other very shade-tolerant species—eastern hemlock (chapter 3) and American beech (chapter 6). Unlike beech, sugar maple does not reproduce clonally through root sprouts, nor does it have specific requirements for establishment as in hemlock. Seeds of early successional trees often accumulate in the soil over time. Whenever and wherever a canopy opening occurs, seeds are ready to germinate and start growing for the canopy. This accumulation of dormant seeds is known as a seed bank. As a classic late-successional tree, sugar maple does not form a seed bank but rather a seedling bank—a pool of suppressed seedlings that stand waiting until an opportunity comes along that allows growth. In this way, the sugar maple does not have to wait for seeds to be produced in the fall and germination to occur the following spring before taking advantage of a canopy opening. Instead, their head start allows them to grow immediately from their advanced state of development.

The shade tolerance of sugar maple is a necessity for persistence in dark forest understories. The carpet of maple seedlings may be at, or very close to, their ability to break even photosynthetically. The bigger a plant gets the more tissues that are alive and need to be maintained energetically. The only source for the carbon-based sugars that a plant burns during respiration is photosynthesis. So, the larger the seedling, the greater its carbon demands, the more the products of photosynthesis go toward maintaining current tissues rather than growing new ones. Energetically, maple seedlings stall at about 1 foot in height. A seedling this size can be 5 or 10 years old, or perhaps 50, or even 100 years or more, doing just enough photosynthesis each year to make a new flush of leaves and patiently wait one additional year.

The ascent of long-waiting seedlings to the forest canopy is just as slow. When a canopy opening occurs, the sugar maple seedling will have a burst of growth. However, the shade tolerance of sugar maple also constrains seedlings to grow slowly even under ideal conditions. Sugar maple seedlings often grow in mixtures with other tree species, nearly any of which grow faster than the maples. As a result, one opening is rarely enough to allow a sugar maple to make it into the canopy. Typically, multiple canopy openings are necessary before a seedling on the forest floor becomes a canopy individual, persisting there for centuries. After the first opportunity, the maple seedling may be 5 or so feet (1.5 m) tall before it is overtopped by surrounding saplings, stalling growth. Perhaps another 40 years pass before that early winning tree is damaged in a storm, taking out half its canopy. The sugar maple grows more, becoming a few inches in diameter, 20 or so feet (6 m) tall. The canopy tree finally succumbs, 20 years later, to the injuries it sustained in the storm, and the maple pushes forward again; this time, a small portion of the canopy reaches full sunlight. The tree grows slowly with at least some access to full sunlight, gradually pushing back on the surrounding trees as maple can continue to grow even in their shade. As trees growing adjacent to the subordinate maple die or are damaged, the sugar maple will expand, eventually becoming a full canopy dominant tree, perhaps 150 years or more after the seed first germinated.

Sugar maple's growth rings record the cycles of suppression and release. Maples produce microscopic rings in the seedling bank, wider rings during release periods, narrower during suppression. As the tree gets larger and larger, the costs of maintaining living tissues increase, reducing the energy for growth, further reducing the size of growth rings. An even more important energy usage develops once the tree makes it into the canopy—reproduction. Allocation to reproduction further reduces energy available for growth, but growth is less critical once in the canopy. The flowers of sugar maple open in the spring a few weeks before the leaves begin to expand. They are wind pollinated, though insects visit the flowers as well. The fruit are winged samaras, produced in pairs, but typically only one will fully develop a seed.

Despite being pollinated early in the spring, the tree does not release mature fruit and seeds until the fall. The muted wind currents under the forest canopy are well suited to the dispersal of sugar maple samaras. The slightest of breezes release the fruit, spinning gently down to the ground— the helicopters of childhood. The autorotation of the fruit generates lift, slowing the rate of the fruit's descent. Any lateral wind currents will push the seed away from the maternal tree. Dispersal distances are moderate but sufficient to generate at least some longer-distance seed dispersal. Of course, most seeds will remain beneath the maternal tree and be subjected to seed predators all winter long. Sugar maple seeds require exposure to cold, moist conditions to break their dormancy, ranging up to 90 days in northern populations, less southward. While that isn't unusual in and of itself, the optimum temperature for *A. saccharum* germination is just barely above freezing. In nature, the seeds of sugar maple often germinate under an insulating blanket of snow to get a head start on the growing season. An earlier start often translates into a longer opportunity for growth in a seedling's critical first year.

## Forest Mesophication and Maple Expansion

It may feel odd to discuss species' expansion in a book on the loss of tree species, but I believe it is essential to place all tree dynamics in an appropriate ecological context. Species ranges and abundances are plastic over time, as environmental and biotic interactions change. Quite simply, sugar maple is expanding in forests formerly dominated by oaks (*Quercus* spp.) and hickories (*Carya* spp.). Oaks and hickories have high light requirements for successful regeneration, and so their recruitment in forests can be limited. The lack of recruitment is exacerbated by the similar age of many oak–hickory forests, brought about through early logging and the loss of American chestnut. In contrast, shade-tolerant sugar maple seedlings do well under these same closed-canopy conditions, allowing them to expand their populations as other tree species decline. Sugar maple may effectively be the only regenerating canopy tree in many areas, setting up some interesting long-term management decisions.

I regularly sample a local area with my plant ecology class as a forest projection exercise. We survey the tree seedlings, saplings, and canopy trees to project future forest composition. The youngest tree seedlings vary somewhat year to year based on weather conditions and which species had good reproductive years but tend to be more diverse than the forest canopy. We regularly find oaks, hickories, cherries (*Prunus serotina*), hackberry (*Celtis occidentalis*), dogwood (*Cornus florida*), redbud (*Cercis canadensis*), white ash (*Fraxinus americana*), and, of course, sugar maple. Many of these trees will never make it to the subcanopy, let alone the canopy, so projecting from this stage is difficult but generates a good class discussion. The forest canopy of this area reflects the disturbance history of the site—the land was cleared and used as pasture until the Civilian Conservation Corps made the area a state park in the 1930s. The forest retains a few open-grown white oaks (*Quercus alba*) with large lower branches from early growth in the pasture. The rest of the trees all have the narrow crown indicative of trees growing under competition from neighbors. These trees are also diverse and contain several oak species, hickories, white ash, black walnut (*Juglans nigra*), slippery elm (*Ulmus rubra*), and a few sugar maples along a moist ravine. If you add up the area of the trees' trunks, this remains an oak–hickory forest. A dramatic shift occurs when you examine the saplings, the trees poised to next take position in the canopy. These trees are overwhelmingly composed of sugar maple. Most years, this sample contains a few dogwoods too short to make it into the canopy, perhaps an oak or two, and well over one hundred sugar maples. Some years, depending on where the sampling lines are delimited, there have been only sugar maples in the sapling size class. This distribution clearly indicates that sugar maples are the primary seedlings surviving long enough to make it to sapling size. The even-aged nature of most of the stand also suggests that the transition to maple dominance may be rapid as the existing canopy trees age and die.

On the surface, the dynamics of this forest appear to be a normal, perhaps expected, successional transition—more shade-tolerant tree species are replacing less shade-tolerant species. Why, then, is this transition leading to the loss of oak–hickory forests? The answer appears to be fire, or rather the lack of it. For 14,000 years or so, fires were an integral component of many forests in Eastern North America. Fires maintained the grass dominance of prairies and helped to keep the open-canopied structure of oak–hickory forests. More importantly, fires killed thin-barked, susceptible trees, including maples. While fires did occur naturally from lightning strikes, Native Americans set the vast majority for land management. Fires were set to drive game, open the forest understory for hunting, and encourage the tree species that provided the best food resources for wildlife.

Fire frequency dictates vegetation. Prairie grasslands have the most frequent fires and the lowest abundance of trees. A slight decrease in fire frequency results in an oak savannah, where open-grown oak trees are interspersed with herbaceous prairie vegetation. A further decrease in fire frequency leads to oak–hickory forest with its characteristic understory flora. Entirely or almost entirely remove fire to get a mesic forest filled with species of the mixed mesophytic forest, with a different understory flora. The Europeans not only displaced the indigenous people of North America but their forest-maintaining fire regimes as well. As the relentless expansion of Europeans pushed westward, so did fire reductions. Over time, the once widespread and continuous forests were fragmented, disrupting fire transmission. Active fire suppression began in the twentieth century and led to large-scale forest changes. Savannas closed in to become forests, and fire-intolerant trees expanded in forests where they were once rare. Forest dynamics that once occurred over centuries with slow climatic changes were compressed into a single generation of trees.

This change in our forests' composition has been termed "mesophication" by forest ecologist Marc Abrams and his colleagues. It reflects the complex processes that result in a shift from drier forest types toward moister forests. Oaks and hickories do much better with more xeric conditions, the high light and drier conditions characteristic of more open forests. Fire loss allows tree canopies to coalesce, creating the shade that favors more shade-tolerant, mesic species—typically sugar maple and red maple (*Acer rubrum*). These species would often be present at low densities in oak–hickory forests, associated with moister areas in ravines, along streams, and similar conditions that would reduce fire frequency. Denser forests also retain leaf litter moisture, accelerating decomposition and slowing the spread of any fires that do occur. Over time, more mesic tree species spread outward from their fire refugia and compete with the xeric species for canopy space as the whole system shifts toward mesic conditions. The litter deposited by mesic species decays more quickly than does the litter of oaks, hickories, and chestnuts, leaving less to burn, further reducing the intensity and frequency of fire.

A fire or two early in the mesophication process may be sufficient to return the forest to more xeric conditions, but much more intensive management is required as time progresses. Management may be a series of fires in rapid succession, selective logging of mesic trees, or girdling unwanted trees to leave them standing dead in forests. The effects of such practices will accumulate over time as there is no quick fix. One site in the Missouri Ozarks that I visit regularly has recently added fire management, burning three times in the past 10 years. The first fire was spotty but removed much of the accumulated leaf litter and top killed a few of the smaller mesic species (*Acer rubrum* was the mesic culprit there). Resprouting of woody plants was rampant, but some patches of grass and a few of the desired shortleaf pine (*Pinus echinata*) seedlings appeared. The second burn removed a lot more leaf litter, killed some larger trees, and resulted in even more grass cover and pine seedlings. The third fire resulted in large areas of exposed rocky soil, even more trees damaged along the drier ridges, and an excellent cohort of pine and oak seedlings the following year. Fire will need to be reduced for a few years to allow tree regeneration but will need to be a part of long-term site management.

I am deeply disturbed when people discuss mesophication as an aberrant, unprecedented event in forests. There have always been mesic forests, even in the heart of the prairie and oak–hickory forests that border it. In Illinois, an area transitional between prairies and oak–hickory forests, there are several wonderful old-growth mesic forests protected as nature preserves. These forests occur in areas naturally protected from fire even before twentieth century fire suppression. The best of these forests are perhaps the remnants of the Big Grove near Urbana, Illinois, which was historically surrounded by prairie grassland and protected from fire by a curve in the Salt Fork River. The two parcels that remain of the original grove, Brownfield and Trelease Woods, were once dominated by American elm (*Ulmus americana*), sugar maple, white ash, basswood (*Tilia americana*), and oaks, with understories of buckeye (*Aesculus glabra*), ironwood (*Ostrya virginiana*), pawpaw (*Asimina triloba*), and spicebush (*Lindera benzoin*), reminiscent of mixed mesophytic forests. As the

American elms died, sugar maple and other species expanded, but mesic species were always abundant in the forest. These mesic forests are beautiful, much less common, but still perfectly natural in all possible meanings of the word.

The way I see it, land managers have two options for their forests. They can fight succession and the mesophication it generates. That approach will entail a long-term commitment to management, both to push composition toward the desired xeric target and to maintain the forests. Forests that burn tend to be more open and, therefore, are susceptible to the many non-native exotic plants that plague our forests (chapter 7). Therefore, management will also need to address those challenges. The other option is to allow mesophication to occur and, with it, the expansion of red or sugar maple, depending on where the forest is. This approach may prevent many invasions but will be a decidedly different forest than was locally dominant for the last several hundred years. Both are valid choices—the right one for a site will depend on the resources available today, as well as the likelihood of continued management. A one-time thinning to allow oak regeneration may be catastrophic if secondary treatments do not follow to cull the invasive species that almost always follow. I have seen this disastrously done more than once in nature preserves. Selective logging to remove sugar maple may be a valuable revenue source for landowners if other management activities accompany it.

## Sugar Maple Declines

Decline is a vague word and is a term applied here precisely because of that quality. A tree decline is a general term for trees that are not healthy. Sometimes a decline leads to tree mortality, sometimes trees recover and are no longer in decline. If we knew what was generating tree health issues, we would simply call it by the culprit's name, rather than refer to it as a decline. Instead, declines tend to involve a revolving suite of antagonistic factors that sometimes interact to generate a decline, but not always. Declines can be maddening from a management perspective but are illustrative of the complexity of contingent interactions involved in forest ecology. What can alter a tree's health? Weather, mammalian and insect herbivores, generalist and specialist pathogens, nutrient availability, other soil conditions, site history, competition with conspecifics, and competition with other species can all be important. Of course, all the problems discussed in this book at one time were considered declines but then rapidly were shifted to an outbreak of insect A or disease B once the cause was attributed.

Forest ecologists have documented sugar maple declines throughout much of the species' range, dating to the late 1950s, when a decline occurred in Wisconsin. In that decline, a drought was followed by an insect outbreak that defoliated tree canopies, followed by an outbreak of the fungus *Armillaria*, which caused declining health and death of trees. In this sugar maple decline, drought generated tree stress, increasing susceptibility to insect herbivores, making the trees even more stressed and susceptible to an opportunistic fungus. A (drought) + B (insects) + C (fungus) = D(eath). So simple, right? Had the insects also been depressed by the drought or the fungal disease not been present, the impacts may have been scarcely noticeable. Since that sugar maple decline, dozens more have occurred. Often these declines have weather linked with an herbivore outbreak and another stressor. Observations that sugar maple declines were more common on unglaciated soils, which tend to have a lower pH, brought the idea of acid deposition to the forefront of stressors investigated.

Acid deposition from the burning of fossil fuels introduces nitrogen and sulfur-based acids throughout the eastern United States and Canada. These acids come down with rainfall—wet deposition to ecosystem ecologists or, conversationally, acid rain. Once in the soil, the protons released from the acids, $H+$ ions, interact with the soil environment. Clay particles in the soil occur as crystalline sheets, with large, negatively charged surfaces that bind with positively charged nutrients, known as cations. Biologically important cations include potassium $(K+)$, calcium $(Ca+)$, and magnesium $(Mg++)$, along with sodium $(Na+)$ in some soils. In neutral soils, these cations would be held loosely, going into and out of solution in the soil until taken up by a plant root, lost as runoff, or otherwise chemically bound. Soils acidified by acid rain possess protons in great abundance. These anthropogenic acids bind tightly to clay particles, displacing the important nutrient cations. Once released, cations may be absorbed by plant roots but more likely get leached out of the soil. Over time, acidified soils lose fertility as the displaced cations are critical for plant growth. Potassium controls stomatal opening and other metabolic processes, calcium is critical to building cell walls, and magnesium is at the heart of every chlorophyll molecule.

Low levels of calcium and magnesium in leaves or soil are most often associated with sugar maple declines. Nutrient losses appear to be predisposing stressors on maples that can combine with other factors such as drought or herbivores to cause a maple decline. Additional stressors over an extended period are still needed to lead to tree dieback. However, if the trees are starting from a nutritionally stressed position, this may increase susceptibility to weather anomalies, herbivores, or pathogens, producing a greater likelihood of a local or regional dieback event.

The complex suite of causal processes makes sugar maple declines challenging to manage. Fertilization or application of agricultural lime has been advocated in some places to ameliorate nutritional limitations, which is a much more straightforward management practice than attempting to control insect herbivores or generalist pathogens. Much of this text would not be necessary if we could reliably predict insect or pathogen outbreaks in natural systems. It is also impossible to control one pest insect without affecting thousands of innocent nontarget insects, many of which may feed on or compete with the target pest. Fungi are even more challenging to deal with as the soil contains thousands of beneficial fungi.

## The Asian Longhorned Beetle—a Generalist Feeder

The Asian longhorned beetle, *Anoplophora glabripennis*, like all longhorned beetles, is a beauty to behold. By far the biggest of the insect pests that we have discussed at 2/3 to 1½ inches (1.7–3.8 cm) in length, the beetle is dark, an almost black charcoal grey, covered with splotches of white across its back. While this color pattern is striking to look at in isolation, it offers good camouflage against the mottled gray and brown tree bark where this insect typically is found. The Asian longhorned beetle, like all the cerambycid beetles, has long antennae that sweep back elegantly past its back, providing its namesake feature. Incidentally, the cerambycid beetles are a diverse group with about 26,000 species worldwide. For this reason, the common name "Asian longhorned" is descriptive of many, many species of beetle. Any new invaders from this group will require much more specific names.

Unlike the other pests discussed in this book, Asian longhorned beetles are more generalist in their feeding. While I have chosen to focus on sugar maple as the central thematic narrative of this chapter, Asian longhorned beetles feed on many maples in its native and introduced ranges. The majority of maples appear to be adequate for *Anoplophora glabripennis* to complete its life cycle. Beyond being a generalist on *Acer* species, Asian longhorned beetles also feed and reproduce successfully on buckeyes and horse-chestnuts (*Aesculus* spp.), birches, hackberries, already beleaguered ashes (*Fraxinus* spp.), sycamores (*Platanus occidentalis*), poplars and cottonwoods (*Populus* spp.), willows (*Salix* spp.), and elms (*Ulmus* spp.). This list is likely not complete but indicates the range of potential impacts in Europe and North America, where the insect has been repeatedly introduced. While the insects are generalists, they still exhibit clear preferences for maples and even preferences for individual species of maples. But then again, preferences are just that and may determine which trees get attacked first but not limit the pest insect's expansion.

The insect's life cycle and feeding action mimic many of the bark feeders we previously discussed, and I will not belabor unnecessarily here. After mating, gravid female beetles chew out an oviposition pocket into the bark of a host tree and lay a single egg onto the living tissues beneath. After the egg hatches, the larva will chew into the underbark, taking a year or more to develop into an adult, which emerges from the tree trunk and flies away. The adults feed on the twigs and foliage of the same tree species, resulting in minor additional damage. A single female beetle can lay up to ninety eggs during her lifetime. More importantly, egg-bearing female beetles can fly 1.2 miles (2 km) or more to establish a new population. Most beetles in a population will move relatively modest distances unless their host trees begin to dwindle. While the life cycle of the Asian longhorned beetle is much less explosive than other insect pests, the reproductive capacity and the potential for movement across the landscape make these insects a severe challenge to control.

Genetic patterns in Asian longhorned beetles have revealed the history of the insect's spread. The Asian longhorned beetle appears to be a native of the Korean peninsula, where it primarily feeds on the maples *Acer mono* and *Acer truncatum*. In Korean forests, the insect is relatively rare, with most trees uninfected, and the insect appears to be completely absent from some forests. The beetle also seems to be limited to edge habitats along streams and roadways, being largely missing from the interiors of forests. Oddly, North American native silver maples (*A. saccharinum*) planted in Korea are heavily targeted by the insects.

From Korea, the species spread into China. Some data suggest that the insect is also native there, though population outbreaks only began in the late 1970s. Early Chinese reforestation efforts resulted in massive, low-diversity forest tracts perfect for the spread of the Asian longhorned beetle. This insect is now one of China's most economically damaging forest pests. Whether the diet range in Asian longhorned beetle is just expressed more clearly in Chinese forests, or the beetle's diet has expanded, more tree species are attacked in China, including elm, willow (*Salix* spp.), and poplar. Regardless, Chinese populations of the beetle are behaving in a much more invasive manner than those within the native Korean range. Invasiveness in one area is an excellent predictor of invasiveness in another. Chinese populations became the source of invasions in North America and Europe. Untreated wood in shipping containers and pallets served as the movement vector. The long larval period of the Asian longhorned beetle allows it to complete its life cycle in transit and emerge in its new home, ready to mate and start a new life for itself.

The first North American populations were discovered in New York City in 1996 and Chicago in 1998—both major ports of entry. Since those initial invasions, additional infestations have appeared in New York, New Jersey, Massachusetts, Ohio, California, Ontario, and likely elsewhere. European countries with invasions include Austria, France, Germany, and Italy. Sadly, genetic data indicate that there has not been a single invasion event but repeated colonizations from multiple Chinese sources. With enhanced vigilance, packing materials that contain Asian longhorned beetles have been intercepted numerous times in receiving ports, preventing additional invasions.

The latitudinal range of the Asian longhorned beetle in China and environmental modeling of its tolerances suggest that the potential North American range is essentially the entire range of eastern deciduous forests. Similarly broad distributions are projected for Europe. This distribution is yet another example of how the biogeographic linkages of North America and Europe with Asia align to ensure the success of a non-native insect. Host tree genera for the Asian longhorned beetle are broadly distributed across all three continents. All three of these continents also have similar environmental conditions, forming little barrier to the colonization and spread of the beetle.

## Killing Trees to Save Trees

Management of the Asian longhorned beetle focuses on eradicating incipient populations before they fully escape and become uncontrollable. Researchers are working to identify plant chemicals that attract Asian longhorned beetles, pheromones that can lure and trap male and female beetles, and parasitoids in Asia and North America as potential biocontrol agents. Other recommendations unrealistically involve maintaining tree and forest health, maintaining tree diversity, and preventing forest fragmentation. This is a feat impossible in North America's fragmented, largely privately owned forests.

Remediation of all Asian longhorned beetle colonizations has roughly been the same. Immediately, quarantines are set in place to prevent the movement of wood products and nursery materials from leaving the area and potentially dispersing the insect to a new location. These quarantine areas are determined based on the location of infected trees and the potential movement distance of the insect. Survey work also begins as soon as beetles are detected, in which colonized trees are identified with the aid of binoculars, bucket trucks, and tree-climbing equipment. Locating infected trees is much easier in urban or suburban settings where tree access is more straightforward. The need is, however, more critical in forested landscapes. Visual inspections cannot possibly locate all infected trees, so these must be repeated year after year. As searchers find infected trees, they are removed, along with any nearby high-risk trees. High-risk individuals are preferred host species located close to infected trees. It also matters how infected the trees are and how many trees are infected, as that provides a rough estimate of how many adult insects will be looking for new host trees.

The process is not over once crews remove the trees, as the logs can still contain living larvae able to complete development in downed wood. Culled trees are chipped into pieces too small to support larval development or burned. The cure represents a massive effort in a landscape. In cities such as Chicago, entire blocks of trees disappeared overnight to eradicate this insect. Elsewhere, small suburban forests have been converted to sparse woodlands. Armies of people wander city streets and forests monitoring trees. This scorched earth approach has worked to date, but the costs are massive. Then again, estimates of the expense of removing and replacing trees if an uncontrolled outbreak occurs are staggering. Those estimates consider only urban trees—minuscule compared with the potential economic and ecological impacts in natural forests.

While it is easy to support removing a dead or dying tree, getting support to remove thousands of healthy trees proactively is a public relations nightmare. For this reason, all eradication efforts have included a massive public outreach component. Engaging the public as watchdogs for additional outbreaks of Asian longhorned beetle has helped eradication efforts in several cities. Municipalities, private companies, and charities have often picked up the tab for replacing urban trees, particularly street trees. In contrast, there are many fewer options for forest landowners who may lose a majority of their forest canopy to eradication efforts. Forestry nursery plants are typically small, 1- to 2-year-old seedlings that will take decades to fill in the forest canopy. Until canopy closure occurs, competition from native and non-native shrubs and vines will need to be managed, or forest recovery will be even slower. The level of damage resulting from eradication efforts in a forest stand can mimic a commercial logging venture, with, of course, none of the financial windfall.

In June 2011, the Asian longhorned beetle was detected in Clermont County, Ohio, outside Cincinnati. This area is a commuters' haven for those, like my cousin, who work in Cincinnati but want a rural life. He and his wife bought acreage where they raise and train sheep-herding dogs. So, they also have sheep, chickens, ducks, and some peacocks— quite idyllic. The surrounding area is forested land, fragmented by farms, homesteads, and roadways. By the spring of 2017, a quarantine area of 61 square miles (158 km²) had been demarcated in Clermont County. This was small compared with other quarantines, such as in Massachusetts (110 square miles; 285 km²) and New York (137 square miles; 355 km²). By October 2020, over 3.4 million trees were surveyed for Asian longhorned beetle infestation in Ohio. These surveys led to the removal and destruction of 20,319 infested trees. Because of the amount of forest and sugar maple density, an additional 89,101 high-risk hosts were removed. A grand total of over 109,000 trees were removed from 61 square miles! The cure has been nearly as bad for the local landscape as an actual, permanent outbreak would have been, albeit much smaller in scope. The efforts have been successful, removing two adjacent areas from quarantine after no infestations were found over several years. However, 56.5 square miles (146 km²) remain under quarantine.

My cousin went to public meetings as the quarantine and eradication efforts began. As would be expected, there was a lot of anger at removing perfectly healthy trees on private lands. Couldn't they spray something to fix the problem? Personal rights and the freedom to choose are continually at the forefront of the American mind, so the eradication plan went against many people's land ethic of "leave my land alone." We see similar conflicts with vaccinations—the significant difference is that in immunization, nonparticipants in the system, the unvaccinated, bear much of the cost of not joining the communal defense. For Asian longhorned beetle, landowners who would not participate could doom an entire continent's trees. Thus, no choice was offered, and that is difficult to swallow for many landowners.

Considering all the money that is spent in eradication and monitoring, it would seem more straightforward to prevent the insect from arriving in the first place. In response to the initial outbreaks of Asian longhorned beetles, APHIS (the U.S. Department of Agriculture's Animal and Plant Health Inspection Service) developed additional regulations in 1998 that now require wood packing materials from China to be certified and fumigated to prevent live insects from entering the country. Rules allow any wooden packaging that does not meet standards to be turned away. Of course, by the time the packaging gets inspected in a port, adult insects may already be flying around the warehouse's interior and dispersing into the surrounding landscape. Further, not every piece of packing material can be inspected, generating additional opportunities for insects to slip by undetected. It is worth noting that the Ohio infestation occurred in 2011, well after the new regulations. Clearly, legal remedies have not been sufficient.

## Maple's Future in North America

As invasions, so far, have been successfully rebuffed, we have little information on what the effect of Asian longhorned beetles might be in North American forests. Asian longhorned beetle infestation in metropolitan areas has killed many host trees, leading to the relatively rapid discovery of this large beetle. Almost by definition, trees in cities and suburbs are edge trees, growing in full sun like the trees preferentially attacked by Korean insect populations. City trees may also be stressed by soil compaction, lack of water, and air pollution, perhaps leading to increased susceptibility to the insect's damage. Asian longhorned beetle has colonized only a few actual forests, so there is little direct information on its effects on natural systems.

Tree mortality appears lower in forest sites than in cities, or perhaps this mortality is just a slower process. The preference of Asian longhorned beetle for maple species is clear, with damage restricted to that genus only. The beetles show a clear preference for red maple over sugar maple and Norway maple (*A. platanoides*), a horridly invasive tree in northeastern forests (chapter 7). The invasive beetle populations were "controlled" quickly, so additional information on natural forests is thankfully not available. Hypothetically, an uncontrolled outbreak could develop in a forest as follows. Roads, developments, and other incursions generate edge habitat that dissect most forests. These edges will provide travel corridors and an environmental base for expansion into intact forests. Initial infestations will occur on preferred *Acer* species. As their population grows, Asian longhorned beetles may shift to forest interiors or less-preferred host trees. Even if we assume little direct maple mortality, their damage will undoubtedly contribute to the potential for regional decline events. Atmospheric deposition has likely stressed many sugar maple trees throughout the eastern United States; add this in combination with weather fluctuations, and the likelihood of appreciable declines and mortality increases dramatically. Outbreaks of Asian longhorned beetles in China are often associated with drought stress, so the potential for weather to enhance the insect's impacts is well established. Even without directly causing mortality, Asian longhorned beetles may add to the plethora of stressors and exacerbate maple declines, leading to mortality and canopy opening.

This scenario is perhaps the best case for our forests under Asian longhorned beetle invasion. It assumes that the preferences of the insect are strong enough to limit their spread. It also assumes that the rapid mortality of host trees seen in urban areas will be reduced in continuous forests. Both assumptions are unlikely to be borne out in natural systems. The invasiveness of the beetle in China's forests argues that spread and damage will be much greater. I have also primarily considered the scenario of this beetle as a pest of *Acer* species rather than what might happen if the Asian longhorned beetle infected the full spectrum of trees that can be its host. On some level, the dramatic decreases in elm from Dutch elm disease and ash from the emerald ash borer have already reduced hosts of this species and its potential forest impacts. The most widespread Asian longhorned beetle hosts not otherwise affected by pests are birch and poplar. Most trees affected by Asian longhorned beetles occur in mesic or riparian habitats, especially willow, sycamore, and hackberry, in addition to those listed above. The distribution of these host trees suggests that these mesic to wet forests may be the most affected, as found in Korea. I suppose Asian longhorned beetle feeding would counteract forest mesophication, pushing the balance back toward oaks and hickories in forests traditionally dominated by those species. That, however, is not much of a silver lining.

# 6

# OTHER TREES WITH
# OTHER CHALLENGES

In a text such as this, it can be challenging to select focal tree species or forest pests. Full treatments of all tree species that have been affected by introduced pests and pathogens would be encyclopedic and run the risk of presenting an overwhelming challenge for us as a society to face, which would counter my motivations for this book. Similarly, focusing on only a few trees may generate the false impression that there are few problems. The reality is that there are relatively few canopy trees that aren't challenged by at least one aspect of their biology, whether it is regenerating under deer browse (chapter 7), a new pathogen, or climate change. The purpose of this chapter is to quickly address a few other tree species that I think warrant attention. Each tree has a slightly different challenge it faces, which highlights ecological aspects that are worth discussing. I have left off other forest pests that are also of concern. Oak wilt (*Ceratocystis fagacearum*) and white-pine blister rust (*Cronartium ribicola*) would likely be the next two diseases I would add, but I had to draw the line somewhere. As in the previous chapters, I will focus on the tree's ecology and their threats, just with briefer commentary.

## Flowering dogwood (*Cornus florida*)

CORNUS FLORIDA , L.

Flowers and young leaves of flowering dogwood, *Cornus florida*.

## Dogwood's Ecological Roles

One of the major seasonal events of spring is the blooming of the flowering dogwood, spread in white layers throughout deciduous forests of the eastern United States. Ringed by four leaves that have taken over the visual role of petals, each dogwood "flower" contains several individual flowers that bloom inconspicuously at the center. These individual flowers reproductively mature over time as the large aggregate flower continually markets the presence of pollen and nectar to passing pollinators. Dogwood flowers are, hands down, the most spectacular of the spring-flowering trees, rivaled only by the much less broadly distributed magnolias. *Cornus florida* was nearly ubiquitous in deciduous forests, commonly forming a nearly solid understory of white blooms in spring and red fruits in the fall. The commonness of flowering dogwood means that it played a prominent role in forests—the sheer biomass it represented ensured that. As the only understory tree discussed in this book, it may seem somewhat odd to include it here. But the broad geographic spread and its dominance in its ecological niche strongly argue for the species' inclusion.

Flowering dogwood plays two distinct roles in forests, though these are perhaps not mutually exclusive. They form a large part of the understory layer in mature forests and are also key early colonizing species in disturbed habitats. As a forest understory tree, the species depends on canopy openings for regeneration, persisting long after the gap has closed in the shady understory environment. Flowering dogwoods are capable of rapid growth when in high light, a common characteristic in a gap-regenerating species. For old-growth forests, flowering dogwood stands would have historically been exposed to recurrent fires, which may have also encouraged dogwood regeneration. The contemporary reduction of fire in most forested systems and the denser forests that develop without fire appear to have decreased the abundance of flowering dogwood.

Early after the abandonment of an agricultural field, flowering dogwood can rapidly colonize and grow to fill the area as one of the earlier trees to compete with herbaceous perennials. The fruits of dogwood are bird dispersed, and the seeds contain multiple, thick-walled chambers that protect an embryo from the small mammal seed predators common in early successional environments. The dogwoods may keep pace, at least initially, with the growth of early successional canopy trees. However, as a shorter understory tree, flowering dogwood quickly shifts from rapid growth to reproduction. As these first trees begin to flower, they attract

birds that may bring other seeds into the young forest but, more importantly, take the dogwood's seeds elsewhere. The taller species continue to
grow upward, forming the first true canopy of the young forest, leaving
C. *florida* below.

The fruit of flowering dogwood represents one of the species' more
unique contributions to the local ecology of birds, specifically Neotropical
migrants. As the dogwood fruit ripens, it turns a brilliant red, lovely in
the garden and highly visible to passing birds. The fruit of C. *florida* is a
high-quality fall fruit, in contrast to many others that offer only limited
resources to migrating birds. Dogwood fruit is packed with high concentrations of energy-rich lipids and body-rebuilding proteins—just what a
migrating bird needs on the long trek to wintering grounds. So why aren't
all bird-dispersed fruit similarly provisioned? There is a substantial trade-
off here. Fruit high in lipids spoil quickly—think of how quickly an avo-
cado, another fat-rich fruit, goes from unripe to intolerably past in your
pantry. Dispersers ignore the fruit once spoiled, leaving the seeds to drop,
undispersed. The alternative strategy is to make lower-nutrition fruit that
remain unspoiled for much longer. Such fruit may not be preferred but
will be available when other foods are long gone during winter. These two
options form the extreme ends of the seed disperser reward spectrum for
birds, though there seem to be remarkably few species in the middle.

An additional problem for plants that depend on migrant birds for
dispersal is the ability of the birds to see the fruit. Bright red fruit stands
out in a forest, but how visible are they to a flock of birds passing high
overhead? Many plants on the high-quality end of the fruit spectrum also
signal passing birds with their foliage. Everyone is familiar with foliage's
fall color change that signals the end of the physiological year. However,
some woody plants stop late-season photosynthesis before physiologically
necessary, instead presenting an earlier and more prominent signal to passing migratory birds. In flowering dogwood, this means all the leaves also
turn a deep red. Other trees producing high-quality fruit for migratory
birds, such as sassafras (*Sassafras albidum*—turns orange) and black gum
(*Nyssa sylvatica*—turns brilliant red), also use their foliage to produce a
large-scale avian signal.

The cost of investing in high-quality fruit appears if the timing of ripening misses the timing of fall bird migration. If this happens, there is
a strong chance that the fruit will spoil undispersed. I have often seen

dogwood fruit on garden trees blacken and fall to the ground. With such a cost, what makes stopping photosynthesis early and allocating a lot of resources to the developing fruit worthwhile? Migrating birds are insatiable, stopping to gorge on fruit for a day or two to regain body mass before they take off again. Entire populations of fruit may be consumed, spreading seeds wherever the birds move in those few days, leaving little fruit and seeds to waste.

## The Dogwood Disease

Dogwood anthracnose first appeared nearly simultaneously on both the east (1978) and west (1976) coasts of North America, infecting both the flowering dogwood (*C. florida*) and the western congener, Pacific dogwood (*C. nuttalli*). The progression of the disease is straightforward; necrotic (dead) spots first appear on the tips of leaves, typically lower in the canopy. Next, the infection migrates upward, eventually entering the stem and becoming systemic in the tree's permanent tissues. In the spring, the fungus produces masses of spores from small cankers on infected dogwood twigs; the spores are splashed about by spring rains. As birds often perch on understory trees, they are likely responsible for larger-scale, regional movements. As the disease progresses, it moves farther and farther down the stem, producing larger cankers, damaging branches, and eventually girdling the stem to kill the tree. Death can occur in a few years for established trees, much quicker for smaller plants.

As such visible and valued parts of the forest flora, the disease was noticed quickly, and research began to identify the anthracnose culprit. A fungus was rapidly isolated, but it eluded identification. The responsible fungus is a coelomycete, an imperfect fungus for which sexual reproduction is unknown. As fungal identification is often based on the structures associated with sexual reproduction, identification is tricky for those that only asexually reproduce. It wasn't until 1991 that the fungus was determined to be a new species, *Discula destructiva*. The fungus is thought to be an invader from Asia, but this conjecture has not been verified.

The arguments for the fungus being non-native are as follows. First, the simultaneous appearance of anthracnose on both coasts suggests an outside source rather than the spontaneous evolution of a new pathogen. Second, the diversity of invading populations is often low because of the limited numbers of individuals that colonize a new area. Diversity in *D. destructiva* is low across North America, much lower than that of native fungal relatives. Finally, the Kousa dogwood imported from Asia, *C. kousa*, is resistant to dogwood anthracnose, suggesting an Asian origin for the pathogen, or at least an evolutionary history with *C. kousa*. The prevailing thought is that the fungus is a minor component of the fungal flora on Kousa dogwood, one that had escaped notice until it jumped to naïve North American species.

Identifying the species and potential origin of the fungus are moral victories at best, as the disease has now colonized much of flowering dogwood's native habitat. Disease progression also reflects the duality of the ecological roles of flowering dogwood—as a colonizing tree of open environments and as an understory tree of mature forests. As a component of cool and shady forest understories, anthracnose has hit dogwood particularly hard. The spread of anthracnose within a dogwood population requires the tree-to-tree movement of spores. The tiny spores need to germinate and penetrate a leaf to infect a tree successfully. Even for such a destructive pest, the transition between spore and entering a leaf is a very tenuous process dependent on not drying out. For this reason, cool forest understories where moisture conditions allow fungal growth facilitate disease progression. The fungus typically colonizes lower branches first, pruning the tree upward as canker development kills branches. Spread in forest habitats can be rapid, killing more than 75% of a dogwood population within 10 years or so.

In contrast, dogwood anthracnose spread and damage is much lower in early successional forests or forests with open canopies. These habitats typically have much greater sunlight penetrating the canopy, drying off leaves more quickly and preventing the success of dispersed fungal spores. This is not to say that the trees in open habitats are protected from the disease, but the incidence is much less and low enough to allow tree regeneration. While most studies have focused on highly susceptible forest understory populations in closed-canopy forests, those that have focused on more open habitats have documented dogwood population growth over time. While cutting down forests to allow flowering dogwood regeneration seems ludicrous, simply reintroducing a fire regime into a forest may sufficiently open the canopy to slow the disease's spread and will enable dogwood regeneration, which is largely absent from undisturbed forests. This represents yet another way that land managers can successfully employ fire in forests.

I see the aftermath of dogwood anthracnose in one of my favorite Ozark haunts. The forests are undergoing some level of mesophication, with red maple (*Acer rubrum*) growing from the stream bottoms, where it belongs, up to the ridge, where it does not. The property had an excellent understory of dogwood, and within recent memory, the flowering has declined dramatically. Under shortleaf pines (*Pinus echinata*) and other open areas, large dogwoods persist healthily. Under the denser forest canopy—predominantly resprouted trees following the last round of logging a few decades previous—there are large, dead dogwood stems. The disease was recent enough that most of these stems still stand as sentinels testifying how beautiful this forest was just a few years ago.

## American Beech (*Fagus grandifolia*)

### The Beauty and Use of Beech

The American beech is one of my favorite trees. Besides being an integral component of beech–maple, northern hardwoods, and other forests, they are simply beautiful trees. If they weren't so abysmally slow growing, I'd have planted dozens by now. The single most unique thing about a beech tree is its bark. Medium grey, it is incredibly smooth across the whole length of the tree, from the largest trunk to the smallest branch. Unfortunately, the protective layer of bark is also thin, making beeches quite susceptible to even the slightest fire. With the large sizes that beeches can attain, the trees take on a genuinely regal stature in the forest. The smooth bark surface is irresistible to young lovers who carve their initials into the trunk. Owing to the fickleness of love, you might also see initials crossed out and replaced by others. It is a horrible way to mar a tree, the tree equivalent of a regrettable tattoo.

Plants in the genus *Fagus*, like all species in the Fagaceae (a plant family named after beeches), produce large seeds enclosed in an outer covering. In the oaks, this outer layer takes the form of the acorn's cap. In contrast, the chestnuts (*Castanea*, chapter 2), the chinkapins (*Castanopsis*), and beeches produce multiple seeds wholly enclosed in a spiny burr. Though

FAGUS AMERICANA, Sweet.

Leaves, flowers, fruit, and seedling of American beech,
*Fagus grandifolia* (formerly *F. americana*).

the beeches have among the smallest seeds of the trees in the family, they are a significant resource for birds and mammals. Beeches often occur in areas where the larger-seeded species are less common—deeply shaded mesic forests, in particular. The American beech makes up for the small seed size by producing many of them, providing a ready fall food source for wildlife and, historically, people.

Because of their slow growth, especially when they occur in a thick forest, beech trees have very hard, dense wood. This trait has made beech a favorite wood for producing woodworking tools in the Old World (*Fagus sylvatica*) and the New World. Carpenters traditionally constructed the bodies of hand and molding planes from beech because of its ability to resist wear when being pushed over workpieces thousands upon thousands of times. Occasionally, you will find a finely detailed molding plane with a small strip of boxwood (*Buxus sempervirens*) that forms a particularly delicate portion of the profile. Boxwood is even denser and more rigid than beech but entirely too uncommon to produce the number of hand planes that woodworkers traditionally used. Wood-bodied planes were the norm for hundreds of years, replaced briefly by steel, but now almost entirely supplanted by the router and other motorized tools. Nevertheless, many older wood planes remain functional, proudly retaining the marks of who made them and sometimes who owned them. There is nothing quite like the swish of a hand plane as it satisfyingly forms a long peel of wood.

## Beech Bark Disease—an Unlikely Collaboration

The problem for American beech started in the 1890s, with the importation of ornamental beeches from Europe (*F. sylvatica* and varieties). The beech scale insect, *Cryptococcus fagisuga*, arrived on these trees. Scale insects are sucking insects that use piercing mouthparts to feed on plant juices and can be quite destructive, despite leaving minimal visible damage. *Cryptococcus* loosely translates to hidden balls, so the insects aren't much to look at. The thin bark of the beech comes into play here in a sad way. Most sucking insects feed on foliage or young stems that haven't yet developed the protective cork layers impenetrable to such a weak insect. The thin bark of the beech allows the scale insect to feed directly on the living tissues beneath mature bark. The insect does not hide in the forest canopy; instead, it brazenly lives on the main trunk near ground level. As the insect feeds, it produces a waxy layer that protects the insect from its own, would-be predators. This waxy coating, as it accumulates, is typically the most visible aspect of infection. Unfortunately, this coating also protects the insects from most pesticides, making them very difficult to control. Beech scale insects reproduce only through the parthenogenetic (asexual) production of eggs, so one insect is sufficient to start an infestation. The origin of these insects is not completely clear, though they may be from the area around the Black Sea.

For a while, as the visible manifestation of beech decline, the scale insects were blamed for the beech trees' death. It wasn't until much later that an accompanying fungus was discovered to be the true culprit. Repeated feeding damage by beech scale insects generates opportunities for fungi to penetrate the protective bark and infect the underlying living tissues. Beech bark disease is now known to be caused by *Neonectria* fungi—ascomycete pathogens. Interestingly, the condition is caused by two species of *Neonectria*, *N. ditissima*, a native generalist pathogen, and *N. faginata*, a non-native fungus. *Neonectria faginata* is considered the primary cause of the disease, but the native fungus may cause disease in some areas. As with other examples already presented, environmental stresses may lead to outbreaks of insects and disease. However, the low pathogenicity of the fungus–insect duo generates a disease that has slowly spread for over one hundred years.

At the advancing front of beech bark disease, scale insects are at low abundance, and *Neonectria* infection is almost nonexistent. As time goes on, *Cryptococcus* populations increase, colonizing more trees within the forest. The insects aren't good dispersers, so they hitch rides on birds and mammals to find new hosts. After sufficient insect densities develop, and likely with the assistance of weather-induced plant stress, an outbreak of both the beech scale and a *Neonectria* fungus occurs, generating a killing front. Following sufficient damage, the tree stem is girdled and dies.

It is not the entire plant that dies, just the infected trunk. One of the ecological attributes that beeches possess for low-light regeneration is the ability to spread clonally via root sprouts. These "sons of beeches" (apologies for the obligatory forestry pun) are common in forests, spreading without the dangers associated with reproduction via seeds. After the main trunk dies, the remaining energy produces a thicket of new, young beech shoots. Many of these are a few yards or so from the original tree. As the wave of mortality passes and insect and pathogen populations decline, beech sprouts grow, and the forest recovers for at least a little while. However, the slow growth rate of beech in forests means that it will be decades before beech resumes its former stature in the forest, assuming the disease does not return.

A boost to soil nutrients may also contribute to the demise of beech. While we usually focus on the acidic consequences of acid rain, other chemicals also compose the acids; in particular, sulfur and nitrogen are nutrient elements from fossil fuels and the plant bodies that initially formed them. Once acids release their protons (hydrogen) into the soil solution, the nutrients remain. While sulfur is rarely limiting, nitrogen is the single nutrient that is limiting in most terrestrial ecosystems. As the critical component for the enzymes and proteins involved in nearly every biological process, many organisms spend resources searching out nitrogen rather than energy, as we often think. Acid deposition is necessarily linked with nitrogen deposition—fertilizer from heaven, as it were. Beech scale insects, like any insect, depend on acquiring sufficient nitrogen for growth and reproduction. Infestations on older beech trees are often more successful because there is a greater nitrogen concentration in the bark. With additional nitrogen from rainfall supplementing local nitrogen pools in the soil, beech bark disease may increase in severity over time as stands become increasingly loaded with nitrogen. It seems odd that improving a tree's nutrient status would be detrimental, but it can be.

## Black Walnut (*Juglans nigra*)

### The Use and Ecology of Black Walnut

Black walnut wood is one of the most beautiful woods that the eastern temperate forests of North America produce. A close second is that of butternut, *J. cinerea*, famed for its ability to produce incredibly detailed wood sculptures in the hands of a master carver. Western species of *Juglans* are also valued for their wood. Of course, people also consume walnuts, whether it be the easily crackable English walnuts (*J. regia*) or the more full-flavored black walnuts that take an incredible amount of work

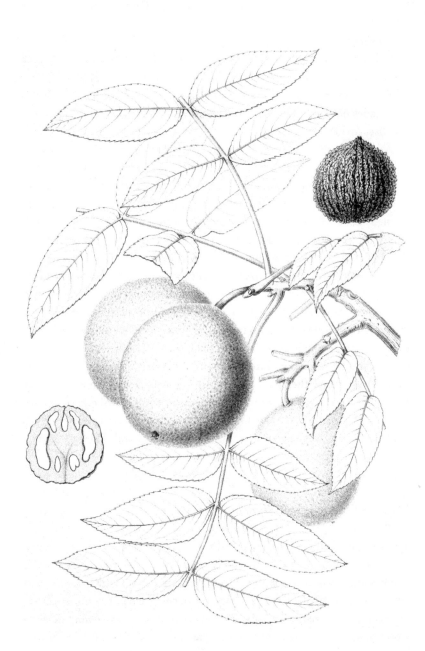

JUGLANS NIGRA. L.

Leaves, fruit, and seed of black walnut, *Juglans nigra*.

to extract. Walnuts are valuable resources for foraging mammals, particularly squirrels. Dozens of black walnuts are carried across the road and planted into our front lawn and garden each year by squirrels attempting to outwit their neighbors. The oils contained within a walnut, besides providing flavor, are energy dense and inspire would-be seed predators to tempt fate by crossing a highway. If they return to eat the buried treasure, they call every predator in the neighborhood as they crunch through the walnut's concrete-like outer wall. Squirrel hunters, human and otherwise, focus on that sound in autumn forests as a tell-tale sign of where their prey sits in the forest canopy.

I must admit that the lumber from black walnut has been the primary driver of my love affair with the tree. The heartwood has a deep, chocolate brown color, rivaling, if not surpassing, the beauty of nearly all tropical hardwoods. Rural legends exist of single large logs ideal for producing veneer valued at $70,000 or more. This fabled value has led many farmers across the Midwest to reserve an acre or two from crops to plant orchards for their grandchildren's benefit. They plant the black walnuts in long rows tight enough so that the growing trees shade each other, causing the trees to grow tall, straight, and drop their lower branches so that the tree bole is ideal for veneer making. The trees need to be much larger than those of other timber production schemes as it is only the heartwood that possesses the beautiful dark color. Larger trees ensure enough of the valued heartwood to make harvest worthwhile—which may require a time span of 40–50 years or more on good sites.

Several years ago, I went to a local auction of the estate of a man who had run a personal sawmill. I think it was officially a commercial enterprise, but it appeared to be an extensive and expensive hobby. After going through wagonload after wagonload of woodworking tools, the auctioneers moved to the pasture to auction off lumber. My compatriots and I had arrived earlier in the day to examine the lumber; it was mostly oak, but there were a few stacks of cherry (*Prunus serotina*), with its reddish-orange heartwood. I had a pocketknife and checked the lower levels of several stacks. In one pile, my pocketknife revealed a stockpile of black walnut camouflaged under a half-dozen layers of cherry boards. After hiding that board, we noted the pile number and walked away casually. The wood was auctioned by the piece—we got the whole pile for $0.65 per board and made off like bandits. I have written a good portion of this book on a walnut desk made from that lumber, atop a cherry writing stand, also from that purchase. It is decadent and beautiful, and all mine.

One of the fascinating things about black walnut trees is how they compete with other plants. Black walnut trees typically occur on rich mesic soils, often in floodplains or similar habitats. They grow rapidly in full sun, making them excellent colonizers of open habitats, although their large seeds limit their ability to disperse. Unfortunately, floodplains are commonly choked with herbaceous plants that will compete with the young tree. In defense against competitors, walnut trees produce allelochemicals—chemicals deployed to inhibit competing vegetation in a process known as allelopathy. This defense operates much the same as plants that produce chemical defenses against herbivores. Named after the genus for walnut, juglone is the chemical that inhibits the germination and growth of many surrounding plants. Walnuts actually produce hydrojuglone, which rapidly oxidizes to juglone in the soil. In many ways, juglone has become the gold standard for allelopathy studies, used to determine how toxic a plant extract is relative to the walnut baseline.

## Thousand Cankers Disease, Natives Out of Balance

The challenge facing black walnut is a bit different than the others described in this book for the simple fact that no nefarious foreign agents are responsible but rather a within-continent range shift. The players are familiar enough, a beetle and a fungus. The walnut twig beetle (*Pityophthorus juglandis*), native to the southwestern United States, is a small (5/64 inch; 0.2 cm), scolytid beetle (the bark beetles) that feeds on walnuts, particularly *J. major*, a native of Arizona and New Mexico, and *J. californica*, of obvious origin. As with other bark beetles, they burrow into trees to lay eggs, and their larvae consume living tissues under the bark. Along the way, the beetles pick up fungal spores, spreading them from tree to tree. One of the many associated fungi, *Geosmithia morbida*, is yet another ascomycete fungus that has lost the ability to reproduce sexually. Neither the fungus, nor the walnut twig beetle, have any significant effect on native southwestern walnuts, which all coevolved together. Had we left well enough alone, this association might have never come to light.

However, the prized economic value for black walnut wood led many people to plant it well outside its native range, spreading the species to the

American West. We can debate the wisdom of planting a tree with clearly mesic moisture requirements into landscapes famed for lack of water, but that is a discussion for another venue. With sufficient irrigation, *J. nigra* can thrive in the arid West for a long time, producing nuts and timber. Whether it was timber movement from a southwestern native walnut or planting black walnut close to native stands of southwestern walnuts that started the disease is unclear. However, the eastern walnut species met up with the native western beetle, who carried a likely native western fungus, and created an epidemic. The resulting disease became ominously known as thousand cankers disease for the multitude of small cankers in the bark that the fungus, in concert with the beetles, produce on infected trees. Though all species of *Juglans* are somewhat susceptible to the disease, black walnut is by far the most vulnerable species.

Early in the disease's progression, the foliage turns yellow and smaller branches begin to die, resulting in a thinning of the walnut's canopy. As the disease produces more and more cankers, larger branches become girdled and die. Branch death may result in the characteristic formation of clusters of shoots just below the lowest canker. Eventually, sufficient damage occurs that the tree dies. Co-occurrence of thousand cankers disease with opportunistic fungi and insects made the initial determination of the disease/vector combination challenging on dead trees.

Early in an infected site, there may be few symptoms to note—a small number of cankers represents trivial damage to established trees. As local insect and pathogen populations increase, damage will also increase but may require a little push from something else. Water stress or other sources of damage can make plants predisposed to pathogens and beetles, much like we have seen with tree declines. Such an inconspicuous initial stage makes quarantine efforts very tricky to manage. The disease may have been present in some Eastern forests for 10 years before detection. The beetles move little, so pheromone-baited traps do not locate the insect unless placed very close to an infected tree.

The key to dispersal for thousand cankers disease is the movement of infected wood rather than insect movements, bringing up another important detail. Genetic analyses of pathogens often focus on reconstructing origins and movement pathways. For the pathogenic fungus, *Geosmithia morbida*, genetic analysis revealed remarkable diversity in the desert southwest, much more diverse than expected for the products of an invasion, which should include a small handful of genotypes at best. This variation strongly argues that the fungus is native to southwestern *Juglans* species. The prolific formation of many small cankers on a tree, each of which may support a different fungal genotype that reproduces asexually, has led to diverse fungal populations in eastern forests as well. Therefore, even a few pieces of firewood or lumber may harbor dozens of cankers and transmit several fungal genotypes to spread into a new area, instantly generating a diverse pathogen population.

The earliest outbreaks of thousand cankers disease occurred in 2001 within western orchards. Under these circumstances, ideal for building large populations of insects and fungi, tree death was rapid and complete. That realization resulted in an immediate and loud outcry and quarantine effort. The quarantine failed, as the disease was already well established in several eastern states within the heart of black walnut's range. Urban and suburban trees, and those along forest edges, have succumbed to the disease, but the disease is rarely found in forest interiors. The disease directly led to mortality in some situations, but in others, the disease stalled, allowing trees to recover. It appears that drought stress exacerbates thousand cankers disease, whereas moist years can allow at least temporary recovery. We simply have too little data to have any confidence in how nonplantation trees will fare long term. The best-case scenario would be that thousand cankers disease generates a walnut decline requiring drought stress to cause tree mortality. As we currently have no way to stop the fungus or its insect vector, we have no remedy to apply if the disease spreads to forest interiors.

## White Oak (*Quercus alba*)

### Oak Ecology

Oaks (*Quercus* spp.) now form the economic and ecological backbone of many forests in Eastern North America. The species-rich genus represents one of our most valuable timber sources, used in furniture, barrel making, and charcoal. It is an excellent firewood, producing heat over long periods. This characteristic also makes it a favorite heat and flavor source in many barbecue traditions. White oak, in particular, is beautiful and

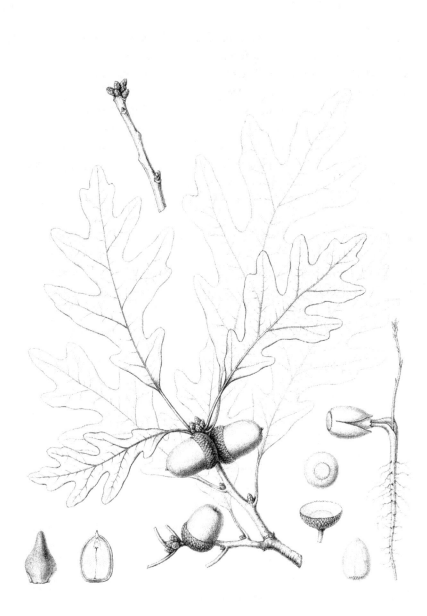

QUERCUS ALBA. L.

Leaves, fruit, and seedling of white oak, *Quercus alba*.

sought out for flooring and furniture making. This species was a favorite medium for Gustav Stickley's furniture designs, central to the American Arts and Crafts movement. The oak love affair continues today with its bold grain prominent in many pieces of furniture.

Oaks are abundant in many forests, achieving high densities in drier upland areas that historically were more prone to periodic fires, lower densities wherever fires were less prevalent. Oaks increased in abundance when American chestnut (chapter 2) died in many forests, replacing that species in much of its range. The area where oaks grow is also where the contemporary removal of fire has shifted the ecological balance in favor of mesic trees, resulting in the previously discussed forest mesophication (chapter 5). Common upland oaks include white oak, red oak (*Q. rubra*), and black oak (*Q. velutina*). Other oak species occupy the wetter side of the moisture gradient, occurring in frequently flooded flatlands near rivers or wetlands. Water-loving oaks include cherrybark oak (*Q. pagoda*), water oak (*Q. nigra*), and pin oak (*Q. palustris*). The Latin meaning of "palustris" is swampy or marshy, so the habitat affinities of pin oak are apparent. The strong habitat selectivity of oaks allows them to grow in different portions of the forest, following their environmental preferences. Different environmental tolerances reinforce the species boundaries among oaks. When they are brought together in a cultivated landscape, hybrids often occur, blurring species lines and identification.

One of the critical roles that oaks play in any forest is in the production of acorns, a food source for small mammals, turkeys, and deer. The fires that Native Americans used to maintain forests encouraged oaks and their game-feeding abilities. We touched briefly on oak reproduction in chapter 2 but will delve into it more deeply here. Like many large-seeded trees globally, oaks do not produce heavy crops of seeds each year but produce large crops irregularly every 3 to 5 years, with occasional crop failure. Such variation in seed production is known as masting. Even more interesting is the synchronization of seed production across individuals. Several mechanisms are responsible for this synchronization, and arguments continue about the most important driver; regardless, it is a fascinating ecological process.

Many plants do not reproduce every year because they do not have enough energy reserves. Flower production is costly; fruit production is even more so if the flowers become pollinated. If it takes 3 years or so to gather enough energy to make a full crop of seeds, reproduction will occur roughly in that time frame. Homeowners often see similar variations in backyard fruit tree production. Limited resources for reproduction explain individual fluctuation but do not generate synchronization across individuals. Oaks are wind pollinated, outcrossing species, so another individual must be flowering nearby for adequate pollination to occur. A plant that flowers out of synchrony will have low levels of pollination, leaving energy reserves for the following year to attempt another bout of reproduction. This effect, known as pollination efficiency, can synchronize neighbors or local populations.

Regional weather patterns can scale up local fluctuations to cover much larger spatial extents. Droughts, late frosts, and other weather patterns affect large areas simultaneously, forming the regional synchronizing force. If a drought occurs late in the growing season, oaks will produce fewer inflorescences to bloom early the following spring. Fewer flowers mean less pollen and lower pollination efficiency, producing fewer acorns the next year. However, low seed set will not deplete the energetic reserves of the oaks, leaving resources in place for the following year. If there are no barriers to reproduction the following year, all the oaks should flower, generate a bumper crop of acorns, and deplete their energy reserves for the next few years. This synchronization will occur at the spatial scale of the drought, causing regional variation in acorn abundances. Unsurprisingly, variation in acorn production has massive implications for the animals that consume them. We will return to this idea shortly.

## The Spongy Moth

A misguided attempt by Étienne Trouvelot to establish silk production during the late 1860s resulted in the introduction of spongy moths (*Lymantria dispar*, formerly called the gypsy moth) into the woods of Medford, Massachusetts. The species spread from this initial location to become one of the most damaging insect pests in North America. The species' current distribution (as of 2020) ranges from Maine to eastern Minnesota and southern areas of adjacent Canadian provinces. Southward, the moth covers the eastern two-thirds of Wisconsin, northern Illinois and Indiana, half of Ohio, and nearly all of the Virginias. Kentucky and North Carolina are poised to be the next U.S. states colonized. Assessment of the environmental tolerances of spongy moths suggests that the spread will continue throughout most of the eastern United States, essentially the entire range of oaks. Although cold limits the population spread of the species, warming from climate change is allowing substantial damage farther northward into Canada.

Horrifyingly, the spongy moth consumes over five hundred species of plants as a generalist herbivore. The species exhibits a clear preference for oaks over other trees, focusing their damage on that genus. However, they can also cause massive damage to fruit trees and several other tree species when at outbreak densities, which they reach at irregular intervals. As eradicating the pest is no longer an option, management to slow the spread now occurs in areas bordering established populations. Combining mapping efforts with species-specific mating disruption has been effective and has halved the rate of spread. Releasing synthetic hormones during the mating season so that males cannot find females disrupts reproduction. To be effective, managers must release hormones over the entire breeding season. The high reproductive output of any successful breeding limits the approach's effectiveness but still slows population growth and spread. Managers may target higher-density spongy moth populations for chemical or biotic pesticides, but these efforts are not specific, killing billions of innocuous native insects.

The life cycle of spongy moths is like that of many forest herbivores. Caterpillars emerge in the spring and disperse by producing a long silken thread to capture wind currents. This mechanism is not particularly good for dispersal, but it is all that the spongy moth possesses. After dispersing, the larvae begin feeding on foliage. As the caterpillars grow, they shift from feeding in the day to feeding at night. During the day, the caterpillars migrate downward to hide along tree trunks or other shaded environments. From this daily migration, the moth acquired the now-pejorative common name "gypsy." (The current common name of these insects, "spongy moths," reflects the texture of their egg masses.) In early summer, the larvae will pupate in the soil or bark and emerge about 2 weeks later. Adult females are approximately 1 inch (2.54 cm) long, males about two-thirds of that size. Females, although they possess wings, do not fly. They crawl up a tree and emit mating pheromones, attracting the motile males to them. After mating, the females lay egg masses on trees, rocks, or other structures, each of which may contain 1200 or more eggs. Female spongy moths lay egg masses on any available structure, including vehicles, contributing to some of their longer dispersal events.

Populations of spongy moths undergo massive population fluctuations over time, with extended periods of low population densities. These periods are interspersed sporadically with massive outbreaks of insects that defoliate large expanses of trees. As spongy moth feeding concentrates on oaks in many forests, these trees bear the brunt of the damage. Defoliation can directly kill trees that are suppressed in the forest understory or are otherwise unhealthy. However, even healthy trees can succumb to the caterpillars' feeding if the defoliation is extensive enough or if the outbreak lasts for more than one growing season. Even where mortality is light, the energetic costs of defoliation can shift competitive advantages from preferred species such as oaks to less preferred species such as maple (*Acer* spp.) and cherry (*Prunus serotina*). Canopy defoliation can also benefit light-starved seedlings of shade-intolerant trees, allowing them to capitalize on higher light levels reaching the forest floor in the middle of the summer. Such impacts may persist for decades in forests, even following a single spongy moth outbreak.

As the long-term prospects for spongy moth control are not good, we can project their effects as they move south and west. Ozark forests traditionally had large abundances of shortleaf pines, but loggers cleared much of this species by the early twentieth century. With their tolerance of dry conditions and resprouting ability after logging, oaks expanded dramatically following this period. Spongy moth feeding may allow the pines to regain dominance, pushing the forest toward a composition more like that before logging. However, it will also represent a massive rearrangement of local industries that rely on the contemporary abundance of oaks. I would expect similar changes in the southern piedmont region, where pines are the classic early successional tree that colonizes large openings, only to be replaced by oaks that can regenerate in the shade of pines. If spongy moths lower oak densities, either other deciduous trees will fill oaks' role or there will be a generation of widely spaced pine forests, like native pines of the southern coastal plain.

If the prognosis was not glum enough, other potential changes on the horizon may make controlling the spread even more difficult. *Lymantria dispar* is sometimes called the European spongy moth to differentiate it from two Asian subspecies of the insect, *Lymantria dispar asiatica* and

*L. dispar japonica.* These Asian insects are much like the European sub-species established in North America, with one significant exception. The females of the Asian spongy moths can fly. Dispersal of *Lymantria dispar* in North America is currently limited by how far larvae can drift on their silken threads or hitchhike on vehicles. The Asian subspecies' ability to fly will allow active movement of females from site to site and, horrifyingly, the ability to find and select host trees. Should either of these subspecies become established in North America, the invasion will move more quickly, affecting forests in a much shorter time frame. To date, there have been a few local populations of these subspecies that have become established, but management interventions have quickly removed these threats. This success is good, but the fact that the overseas invasion routes remain open does not bode well long term.

## The Connection to Public Health

So far, the theme for oaks and spongy moths has been temporal variation—fluctuations in acorn production and spongy moth densities. Within the complex forest food webs, these are not independent processes but interconnected ones with implications for human health. Variation in acorn production has implications for all the organisms that feed on them, from white-tailed deer (*Odocoileus virginianus*) down to the white-footed mouse (*Peromyscus leucopus*). Seed eaters such as white-footed mice are some of the most abundant small mammals in forests, generating substantial impacts on tree seedling regeneration and forming a critical prey base for many predatory animals. As seeds form the bulk of the white-footed mouse's winter diet, the abundance of acorns determines the population size of the mouse. Mast years, those with high acorn production, generate abundant food resources that increase overwinter survival and early white-footed mouse reproduction. As small acorn crops typically follow large acorn crops, this burst of reproduction is short lived, and white-footed mouse populations decline until the next mast crop of acorns.

Fluctuations in seed predator abundance generate year-to-year variation in tree regeneration, with seedling establishment mostly occurring during mast years when white-footed mouse populations cannot consume all available seeds. Variation in the population size of white-footed mice also affects spongy moths. Moth pupae form an important food resource

for white-footed mice during the summer. In years with low mouse populations, they do not significantly affect spongy moth survival. However,
when in high populations following a mast year, they consume enough
spongy moth larvae to reduce population growth and delay an outbreak.
So, large acorn production results in high white-footed mouse populations, which reduce spongy moth abundances—a neat ecological web.
We can bring this all home by noting that outbreaks of spongy moths
defoliate oaks, reducing acorn production. This structure gives us an
interconnected network that, with weather conditions, generates sporadic
outbreaks of spongy moths and their damage.

While this is an interesting story of interconnectedness, it is not the
whole story. Humans intersect this network in one critical way, Lyme disease. This tick-borne disease is caused by a spirochaete bacteria, a bacte-
rium that looks like a tiny spring. The carrier of this disease is the deer
tick, *Ixodes scapularis*. Unfortunately, the common name "deer tick" has
unfairly implicated the white-tailed deer in Lyme disease. In reality, the
predominant culprits are white-footed mice. *Peromyscus leucopus* is a
competent host for the disease, meaning that it serves as a reservoir for
the disease. Tick larvae or nymphs that bite an infected mouse become
infected themselves and then vector the disease to humans. Deer move the
ticks around and increase their contact with people, but small mammals
are the source of the infection. If we place Lyme disease into our forest
web, oak mast increases disease risk by increasing the density of small
mammal reservoirs. Spongy moths counteract this effect by reducing the
energy available for seed production. Long term, if spongy moths cause
a reduction in oaks, white-footed mice will need to switch to other seed
sources. If these new food sources for white-footed mice produce seeds
more consistently than oaks, Lyme disease risk may become more constant
over time. If the new seed sources support fewer seed predators, perhaps
Lyme disease risk will also decline. There are just too many unknowns to
predict with confidence, other than to say things will change. With Lyme
disease affecting more than 25,000 people each year, some with long-term
consequences, there will be human health implications from spongy moth
expansion.

# 7

# THE NEXT IN LINE

Species losses, particularly rapid losses, are critical forest disturbances that set off a cascade of changes that may ripple throughout the forest community for decades, if not centuries. The loss of trees, either through natural gap formation or widespread death due to a pathogen or herbivore, releases the resources those trees had been consuming—light, water, and soil minerals. The resulting pulse of resource availability provides opportunities for new individuals and perhaps new species to become established, expanding local populations or establishing new populations, leading ultimately to changes in forest composition.

More than westward areas, Eastern North America would be naturally forest covered if it were not for constant human activity. Before European colonization, light-demanding species of open habitats were largely be restricted to forest openings and were much less common than they are today. Such openings might have varied temporally, such as the formation of canopy gaps when a tree died, large-scale blowdowns caused by hurricanes, or forest fires that removed the tree canopy. Forest openings also occurred as permanent landscape features, such as areas where the topsoil was too thin to support trees, often called glades, or areas where fires occurred too frequently to allow tree dominance, as in prairies. To colonize such habitats, species needed to establish in high-light environments, often very competitive environments, with many species attempting to capitalize on the available resources. Species specializing in temporally unpredictable forest gap habitats also needed the ability to disperse into newly disturbed areas or to be able to lie dormant as seeds until a gap appeared overhead, essentially dispersal in time. All of these factors that regulated forest openings are still present today. However, climate change is accelerating the rate of weather-related disturbances in many areas.

Within a disturbed forest area, vegetation recovery (i.e., succession) ushers species through the transition from shade-intolerant, sun-loving species of open environments to shade-tolerant forest plants. We often describe plant species as either early successional (shade intolerant) or late successional (shade tolerant) to describe their ecological position. While this dichotomy is a human abstraction of a gradient from early to late successional, it is worth noting that all species, even shade-tolerant ones, benefit from more light. After all, they are shade tolerant, not shade loving. Most plant species can opportunistically capitalize on a pulse of resources. For example, forest understory shrubs may persist for decades in a shady forest. However, they may only reproduce or establish as seedlings when sufficient light is available to support anything other than maintenance metabolism.

In this text, I have often placed our trees in their successional context. Where a species fits along the early to late-successional gradient helps us to understand their natural role in forest communities. We can, however, also place other forest plant life forms into this scheme. Forests are composed of the canopy trees already discussed and herbaceous plants, shrubs, and lianas (woody vines). We can just as easily separate these species according to their successional position as we have the trees. Blackberries (*Rubus* spp.), shrubby dogwoods (*Cornus* spp.), and sumacs (*Rhus* spp.) are early successional shrubs that disappear quickly once the tree canopy closes. Spicebush (*Lindera benzoin*), maple-leaved viburnum (*Viburnum acerifolium*), and pawpaw (*Asimina triloba*) are all late-successional shrubs that persist in shady forest understories. Similarly, the all-too-familiar poison ivy (*Toxicodendron radicans*) is an early successional liana (woody vine) dominant of open areas, and grapes (*Vitis* spp.) are late-successional lianas that persist well past canopy closure.

While this text has primarily been about non-native pathogens and herbivores, species invasions in North America are not restricted to just those species. Intentionally and unintentionally, we have allowed many non-native plant species to become established in Eastern North America. These species also occupy the full range of successional roles in forests, just as our native flora. I would argue that we now have enough non-native species established in Eastern North America to build a functional, non-native analog to our native forests. Such forests could retain all the same functional groupings of species and perhaps even the same physical structure. However, they would be wholly novel ecosystems with unknown linkages with the insects, birds, and mammals of our native forests. Just as with our native plants, non-native plants can capitalize on the next forest disturbance. In some cases, the natives win out; in many situations, the non-natives succeed, spreading the invasion, using more resources, and affecting a larger area. These plant invaders pose an additional threat to many, if not most, forest communities of Eastern North America

The purpose of this chapter is to focus on particularly pernicious plant invaders selected from each forest life form. These species have proven track records of invasiveness and, likely, impacts on forests of Eastern North America. First, however, we need to address one last contemporary pressure on our forests, deer.

## White-tailed Deer Overpopulation

Some are thrilled when they see white-tailed deer (*Odocoileus virginianus*), delighted by a pair of spotted spring fawns or moved by the inspirational beauty of a leaping buck. Others see deer as a menace, devastating crops, landscaping, and vehicle bumpers. I love seeing fawns with their mothers in the spring and early summer before their juvenile enthusiasm is lost and they become another herbivore among my trees. I also allow hunting on my land to reduce the local population. I still get a lot of deer damage when they get in my vegetable garden (now fenced), when bucks rub their antlers on my trees (now caged), or when they browse on nearly everything they can reach. I spend a lot of time and resources trying to keep the deer at bay, and I live in an area where hunting is prevalent. In areas without an established hunting culture, things can be much, much worse.

Deer populations plummeted in the nineteenth century after widespread deforestation and a lack of hunting regulation, reaching a minimum around 1900. With so few deer, many abandoned hunting, and we lost much of our ability for population control of deer. As we had also extirpated gray wolves (*Canis lupus*) from much of Eastern North America by that time, the only real threat to deer was people with guns, bows, or vehicles. A coyote (*Canis latrans*) can kill a young deer, but we had significantly reduced coyote populations as well. From their population minimum, white-tailed deer populations grew in the twentieth century, unimpeded by their locally extinct natural predators and reduced hunting pressure. By then, the landscape had changed to a matrix of forests and croplands, providing abundant food resources in the summer (croplands) and the winter (forests). Urbanization had also occurred, making hunting impractical in many areas and now often illegal out of safety concerns. Over the years, the deer population has burgeoned to a level unprecedented in Eastern North America's history.

Food webs teach us that two things can control a population, food or enemies. As we essentially removed their predators, deer populations grew to become limited by their food. During the growing season, deer find plenty of resources in croplands, backyard gardens, and landscaping. During the winter, deer naturally aggregate into herds, and their food resources become permanent plants in the landscape—woody plants. Winter foraging is concentrated in forests, where deer consume the buds and bark of young trees and low-growing branches. Even unpreferred plant species get "tasted" enough that they sustain damage. Damage accumulates to the point where a browse line is evident in many forests, representing the height that a deer will regularly reach up to nibble a branch. In heavily affected locations, tree seedlings will be absent from the forest floor. If that level of damage does not decrease, saplings will also eventually be missing from the forest. This process continues through the forest age structure until there is insufficient tree regeneration to replace dead canopy individuals or only deer-resistant species remain.

What does the accumulation of forest damage mean for the deer? Competition. More and more deer scrambling for fewer and fewer food resources results in smaller and smaller deer and more heavily affected forests. Of course, hunters like large male deer with large antlers as trophies. Our hunting regulations often encourage taking more males than females to minimize effects on deer populations—a suspect practice when overpopulation is an issue. Sociologically, deer overpopulation is a fascinating phenomenon. If you are a hunter in an area with a high density of small deer, you may opt to go elsewhere to hunt. People from Pennsylvania regularly come to the Midwest to hunt deer and bring home a trophy. Hunting leases are now a major source of farm income wherever deer are large and reasonably abundant. Interestingly, harvesting large male deer by trophy-seeking hunters represents a selection pressure against dominant individuals, counter to natural selection, which favors more robust deer to collect and defend a herd of females. Persistently high deer densities have also allowed the spread of diseases such as chronic wasting disease, which has decimated some deer populations.

High densities of white-tailed deer represent intense pressures on Eastern forests. Beyond the trees and tree seedlings, they also feed on other vegetation that may compete with any tree seedlings that escape deer browse. Over time, forest understories become dominated by deer-resistant plant species. Avoided plant species may be chemically defended by naturally occurring compounds that make them distasteful or physically defended with thorns or prickles that damage the soft mouths of deer. Deer also move species about by consuming fruits and then depositing seeds in their feces or by transporting seeds on their fur or hooves. Trampling by deer also exposes mineral soil, increasing the ability of small-seeded plants to regenerate. Quite simply, it is difficult to understate the complete impacts of deer on forest communities since it is difficult to fully account for both direct feeding effects and all the indirect ways that deer influence forests.

With deer as a contemporary context for our forests, let us now move our attention to the non-native plants in a position to capitalize on the opportunities offered by the loss of canopy tree species. Botanical gardens and arboreta are filled with non-native species on display (not always a good thing), but these species' significance is not just that they are not native and novel. The species highlighted in this chapter can also become overabundant in forest communities, driving down local diversity, altering community composition, and, in general, degrading the communities within which they occur. I suggest that the term invasive better describes these non-native species. Some people do not like this term, suggesting alternative words based on site-specific data and specified levels of impact. This approach requires much more information than is often available. I prefer a more pragmatic definition—they take over.

## Canopy Trees

Across North America, there are relatively few broadly invasive trees when you compare them with the numbers of herbaceous, shrub, and liana species that are problematic. This may be related to the overall slow life history of most trees or perhaps the history of introductions into North America. Invasive trees, other than the two described below, include the princess tree (*Paulownia tomentosa*) and mimosa (*Albizia julibrissin*), but these appear to be limited in their effects and local dominance. Two tree species are particularly aggressive. Chinese tallow tree (*Triadica sebifera*) and callery pear (*Pyrus calleryana*) have both escaped cultivation and are spreading in some locations. They are rapidly developing into problems and could easily have been included here. The two species selected have much broader distributions and have already established themselves as regional conservation issues.

### A Rapidly Growing Colonizer

The tree of heaven (*Ailanthus altissima*), famously referred to in the 1943 book by Betty Smith, *A Tree Grows in Brooklyn,* is a native of China that was actively introduced into the United States by the U.S. Department of Agriculture. Owing to its wide dissemination by the government and well-intentioned gardeners, this species now occurs throughout the United States and southern Canada. The species is characteristic of urban and suburban woodlots, brownfields, and highway margins. From these early introductions and escapes to wastelands, the species has spread into natural areas and is considered invasive in forests across North America. It is intolerant of shade, placing the tree firmly in the realm of early successional trees.

More than anything else, this is a rapidly growing tree, outpacing all but the fastest-growing native trees, as they grow skyward to a height of

60 feet (18 m) or more. Necessarily, fast growth results in poor-quality wood with little strength and little, if any, practical uses. *Ailanthus* has small, wind-dispersed seeds that allow it to blow across roads and sidewalks into open areas to spread far and wide. Interestingly, the species is dioecious, largely having largely separate male and female trees. This characteristic is somewhat odd for temperate trees and even more so for an invasive species. Having separate sexes means that at least two individuals of the opposite sex need to colonize an area before a population can develop. Getting two individuals of different sexes close enough to each other to reproduce in a new habitat is much less probable than the likelihood of a single, self-compatible tree starting a new population. However, when *Ailanthus* reproduces, it produces a massive number of seeds to disperse into the surrounding habitat. This reproductive output is enough to make up for the low likelihood of initial success. *Ailanthus* can also spread via root sprouts, a form of vegetative reproduction. This trait can make the tree difficult to kill. The tree is resistant to nearly all pollution and poor soil conditions, making it bulletproof in cities and other harsh growing situations.

Besides its rapid growth, *Ailanthus* is chemically well defended. As the tree superficially resembles sumacs (*Rhus* spp.) and black walnut (*Juglans nigra*, chapter 6), I often instruct people to rub the leaves lightly as a diagnostic. If the scent is a foul one resembling rancid peanut butter, then the tree is an *Ailanthus*. Just walking through a patch of seedlings perfumes the air enough to make some people gag. Why this stinky species was ever popular in gardens is beyond me. The chemicals produced in the plant's tissues render it unpalatable to most herbivores, always a benefit. Deer won't touch it, so wherever white-tailed deer limit tree regeneration, *Ailanthus* tends to increase. Insect herbivores largely ignore the tree as well.

One insect, however, appears to prefer this tree. The spotted lanternfly (*Lycorma delicatula*), another invasive pest from Asia, first appeared in Pennsylvania in 2014. This insect, a leafhopper, preferentially feeds on *Ailanthus*. The red patches on the wings make it look like a lantern floating in the air. Like all the hemipterans (the true bugs), the insect feeds by sucking phloem tissues, which probably allows it to avoid the tree's chemical defenses. While the spotted lanternfly is often described as primarily a pest of *Ailanthus*, it feeds on many hosts, including orchard fruits and grapes. The presence of *Ailanthus* in so many forests will likely encourage the lanternfly to spread out into the landscape seeking its preferred hosts. Once *Ailanthus* declines, the insect should spread to alternate hosts, increasing its economic impact.

The chemicals that *Ailanthus* release are active against other plants, inhibiting their growth or germination. This allelopathic interaction tends to be common in weedier species that often compete with other plants. A little chemical inhibition of neighboring plants, coupled with the rapid growth rate of *Ailanthus*, is all that is necessary to ensure success in growing above the surrounding vegetation. Unfortunately, chemical production never ceases, so the chemicals continue to be released and affect the herbaceous understory, limiting the abundance and diversity of plants on the forest floor. The shade intolerance of *Ailanthus* prevents this species from establishing in closed-canopy forests, so its time in the canopy is limited, destined to be replaced by shade-tolerant trees.

## A Super Shade-tolerant Invader

The Norway maple (*Acer platanoides*) is a tree much like our native sugar maple (*Acer saccharum*, chapter 5). Native from central Europe to western Asia, this species was introduced to North America as an ornamental tree. They typically have very dark green leaves, although several red-leaved varieties have become popular in home landscaping. The species is also tolerant of compacted soils and pollution, making them ideal for urban and suburban yards where sugar maple often fails. The trees are relatively free from insect pests, making them easy for homeowners to maintain. At first glance, these trees would seem to be ideal additions to a garden. If only they stayed there.

The fruit of Norway maple is a samara, produced in pairs like all maples. The easiest way to distinguish Norway fruit from the native sugar maple is that the samaras' wings are held almost straight outward, nearly perpendicular to the stem that holds them to the tree. Wings on sugar maple samaras hang downward, parallel to each other. The seeds within Norway maple fruit are large enough to provide the new seedling with a good start wherever they land. Wind dispersal is excellent, as the spinning of the fruit generates more lift than the fruit of sugar maple, slowing the rate of descent more, potentially dispersing farther in the forest understory where winds are greatly muted. This dispersal ability has allowed Norway maple to escape from cultivation into surrounding woodlots. It is one of the most common trees in many cities and towns, wherever plants are left to their own devices. Its commonness as a landscape tree also allows Norway maple to use another dispersal mechanism—vehicle windshields, generating an even greater dispersal potential.

Once germinated, Norway maple is very shade tolerant, even more so than the native sugar maple. This extreme tolerance permits the species to grow under meager light conditions, placing it clearly as a late-successional tree. Even without large-scale disturbances, this species can become established and grow, though gap formation will speed it into the canopy. Once a tree reaches a position in the full sun of the canopy, energy is diverted into reproduction, spreading seeds and seedlings across the forest floor. The extreme shade tolerance also has implications for the architecture of the tree. Greater shade tolerance means that leaves can be photosynthetically rewarding to the tree at lower light levels. Maintaining metabolic function allows lower branches to persist on the tree, gathering any remaining light not captured by higher leaves. The multiple layers of the canopy make a dense tree crown that casts a deep shade.

The shade that Norway maple generates produces most of its effects in invaded forest communities. Very little light penetrates the canopy to the forest floor, leaving few resources for even the most shade-tolerant understory plants. The cover and diversity of understory plants is greatly reduced under Norway maple. Furthermore, this dense shade seems to prevent the regeneration of trees other than Norway maple, shifting the next generation of forest tree composition. The dense shade produced is also problematic for homeowners. While the shade cast by Norway maple is cooling in the summer, large trees can inhibit grass growth resulting in a thin, patchy lawn. By the time this becomes an issue, the trees are quite large, and people are loath to remove them.

An additional quality of Norway maple, likely responsible for its lack of insect herbivores, is its chemical defenses. Latex production occurs in a wide variety of plants and is considered a general insect feeding deterrent. Latex produces a milky sap in plants, most familiar from species such as milkweed (*Asclepias syriaca*, famous as the monarch butterfly caterpillar's food) and holiday poinsettias (*Euphorbia pulcherrima*). Sticky latex deters many insect species, particularly those that chew through leaf tissues, releasing the milky sap onto the feeding insect. Deer also seem to avoid browsing on *A. platanoides*, providing an advantage over unde-fended native trees. However, Norway maple is a host of the Asian long-horned beetle (chapter 5), so if we lose the native maples, we will likely lose this invading tree as well.

## Shrubs

The native shrub layer is not particularly well developed in most forests of Eastern North America. Of the many native understory shrubs, few ever seem to be in abundance in forests. Native shrubs are likely in much lower abundance now than they used to be, perhaps linked to the logging history of our forests. Trees may quickly regenerate following a harvest, but under-story shrubs must wait until an understory develops to be successful. The lack of a well-developed shrub layer leaves that niche open, often await-ing one of the many non-native shrubs to fill that ecological space. There are many invasive shrubs throughout Eastern North America; they often dominate local plant communities and displace native shrubs and trees where they become dense. Nearly all are bird-dispersed, so they spread quickly within landscapes. Many of these, such as privet (*Ligustrum vul-gare*), burning bush (*Euonymus alatus*), and Japanese barberry (*Berberis thunbergii*), are horticultural escapees. Others, such as *Rosa multiflora* and autumn olive (*Elaeagnus umbellata*), were intentionally introduced for erosion control and their alleged wildlife value. These species occupy northern areas of Eastern North America. Other taxa, such as heavenly bamboo (*Nandina domestica*) and Chinese privet (*Ligustrum sinense*), dominate southern latitudes with equal effects on plant communities. Many of these shrubs are still available in garden centers, though one must wonder why that is. Any of these shrubs would provide an interest-ing story, but I chose one problematic genus and a new invader. The con-clusions are much the same regardless of which species I discuss.

### A Bristly Newcomer

Wineberry, *Rubus phoenicolasius*, is a native of eastern Asia introduced to North America in the late 1800s. It now occurs across most of the eastern United States, although its distribution is much spottier than many invasive exotics, at least for now. Its common name reflects the feature it is best known for, its delicious fruit. Like all shrubs in the genus *Rubus*, the blackberries and raspberries, each flower produces an aggregate fruit composed of many individual fruitlets, each with a single seed. This species would be a raspberry in gardener's terms. The supporting structure, the original base of the flower, does not accompany the fruit when picked, producing a hollow berry. In blackberries, the structure comes along when picked, supporting the berry's structure with a solid core. This difference makes raspberries more delicate and difficult to ship, so they are less commonly available unless you grow them yourself. The fruits of wineberry are delicious, and the plants are prolific, almost enough for you to forgive its invasive nature.

Like all plants in the genus, the stems of *R. phoenicolasius* are well defended with prickles. Unlike most, however, wineberry stems are nearly covered with a combination of longer and shorter red bristles that enclose the stems. Most native species have only a few lines of these physical defenses along the stems and leaves. While deer will eat and reduce thorny plants such as the invader *Rosa multiflora* and native *Rubus* species, wineberry is just too well defended for deer to consume. In areas with high deer densities, effectively most of Eastern North America, the invasion appears to be facilitated by deer. Likely, deer are foraging on competing species, allowing the invader to expand in the remaining open space.

As wineberry grows, it forms long arching stems that root when they touch the ground, allowing the plant to crawl along the forest understory slowly. In combination with its prolific fruit production and animal dispersal, the species is effective at short- and long-distance colonization. The invader is quite like the native black raspberry, *R. occidentalis*, in this way. Both species primarily occur at forest edges but tend to thin out in undisturbed forests, only to appear again in forest gaps. However, the native produces much less fruit and produces fewer gracefully arching stems. Wineberry, when in its preferred intermediate sun conditions, produces many more stems. These canes pile up on each other, producing thickets of nearly impenetrable, bristly stems 7 feet (2 m) or more high. This mounding habit is where the noxious tendencies of the species lie. Few other plants persist under dense canopies such as this.

Wineberry depends on large forest gaps to become established as seedlings. This requirement places wineberry on the early successional end of the life history gradient, though nearly all forests shrubs will do better in gaps. The species can persist in the shady forest understory once it is established, at least for a while. Ultimately, shade significantly reduces reproduction so that population persistence may be unlikely in undisturbed forests. I have also noted that the abundance of wineberry seems to be more variable over time than it is with other shrubs, perhaps linked with shade or other environmental conditions.

In addition to the lack of deer browse on wineberry, the species may experience lower levels of insect herbivory than similar native species. This idea, known as the enemy release hypothesis, suggests that invasive species may have moved without their coevolved herbivores and pathogens. Native species, in marked contrast, retain their natural enemies and are damaged by them, reducing growth and reproduction. This idea is often tested by comparing native and non-native species in a single habitat and documenting the levels of damage. Relative to the ecologically similar black raspberry, wineberry receives half the herbivore damage that the native does. It certainly appears that the suite of herbivores in wineberry's introduced range preferentially damages the native, adding to the invader's success. Everything indicates that this species will continue to spread and affect the forests of Eastern North America.

## One Bad Genus

The genus *Lonicera*, the honeysuckles, are wildly invasive in Eastern North America, although we have a few native species of that genus. The following section will deal with a viny member of this genus, but several shrub species have become regional invasives. The invasive bush honeysuckles, all Asian in origin, include Amur honeysuckle (*L. maackii*), Tatarian honeysuckle (*L. tatarica*), Morrow's honeysuckle (*L. morrowii*), and some horticultural varieties (*L.* × *bella*, a hybrid between *L. tatarica* and *L. morrowii*). Amur honeysuckle is prevalent in the Midwest and northeastern states; the others are more prevalent in the north-central states, heading into Canada. All bush honeysuckles have a similar ecology. They are rapidly growing shrubs that quickly colonize following a disturbance but persist with canopy closure and often form continuous forest understories. Stands of these species can be dense enough that they exclude all other plant species and severely inhibit tree regeneration.

One noticeable trait of the bush honeysuckles is their pattern of leaf production. Like many of the wildflowers of spring, the shrubs leaf out well before the tree canopy does, capturing light energy in this temporal window of a few weeks. Similar things happen at the other end of the growing season. Honeysuckle species maintain green leaves often until the first hard freeze kills them, capturing another seasonal window of resources when the tree canopy is open. Many think this extended period of leaves allows honeysuckle shrubs to be successful and persist past canopy closure. Early leaf out has other effects on the system. Early nesting birds in understory shrubs or trees are often drawn to the leafed-out honeysuckle canopies to hide their nests. While these locations initially look promising, the bird nests end up too near the ground and are more prone to attack by nest predators than those produced in taller, native cover. Ultimately the presence of bush honeysuckles can indirectly limit bird reproduction—not good for the forest or the birds.

The bush honeysuckles produce red, bird-dispersed fruit that contain lots of water and not much in the way of calories or nutrients. Most honeysuckles, Amur honeysuckle included, ripen late in the growing season, with the fruit persisting into the fall and winter months. The poor nutrient quality means that migrating birds looking to gain weight and rapidly resume their travels ignore these fruit. The poor nutrient quality also means that the fruit do not spoil, persisting on the shrub for a long time. They remain until the winter-resident birds run out of other foods, turning to honeysuckle fruit as a last resort. Beggars cannot be choosers, so the shrubs acquire dispersal services for little contribution on their part. Birds' habit of perching on trees near gaps or along edges for a clear view ensures many honeysuckle seeds will end up in habitats appropriate for establishment.

There may be other indirect effects of bush honeysuckles on forest communities. The dense stands of shrubs that develop can physically exclude deer, allowing tree seedlings to grow unbrowsed. Seedlings under honeysuckle canopies will be in intense competition with the dense but shallow root system. There may be a net benefit of honeysuckle shrubs on tree regeneration in areas where deer pressure is high. Before we call honeysuckle beneficial, I should also note that the dense shrub canopies of honeysuckle, and their high output of seeds, represent safe havens for seed-eating small mammals. These normally wary seed predators are protected from avian predators (owls) and are free to consume all honeysuckle and tree seeds they find under honeysuckle shrubs. The large size of many tree seeds makes them a favorite food item for small mammals. Small mammal populations, or at least their feeding activity, are often higher under shrub canopies, making successful tree regeneration less likely. Without tree regeneration, the shrubs expand as surrounding canopy trees die, potentially further affecting the forest.

There are also direct effects of the shrubs on tree growth. Competition from a dense understory of Amur honeysuckle, and presumably the other honeysuckle species, reduces tree canopy tree growth. As the canopy trees are taller than the shrubs, competition for water or soil nutrients, rather than light, must drive this effect. Slower tree growth means less wood production, delaying the time between wood harvests, reducing profits for landowners. Of course, any tree harvesting disturbs the canopy, potentially allowing bush honeysuckles to expand or colonize a stand, a vicious circle. A new industry centering on honeysuckle and other exotic shrub removal has developed, representing costs to conscientious landowners. The loss of trees such as ash (*Fraxinus*, chapter 4) increases bush honeysuckle abundance as the canopy opens. Such competitive effects are probably common to many non-native shrubs but rarely get investigated.

## Lianas

Woody vines, technically referred to as lianas, are essential components of many forests, from temperate to tropical climates. By many measures, their populations are expanding as forests become increasingly disturbed by human activities. The recent success of lianas is linked to their aggressive life history, somewhat between a woody plant and an herbaceous one. Although they produce woody stems, lianas are structural parasites, depending on another species for physical support. A liana stem needs only to conduct materials up and down the stem rather than physically supporting itself. This freedom allows a liana to allocate more resources to growth, benefiting from permanent plant tissues, but with much less up-front costs to constructing them. Of woody plants, lianas have some of the most impressive growth rates. Not surprisingly then, there is a myriad of invasive, non-native lianas in Eastern North America. Even native grapes (*Vitis* spp.) can behave as invasive in some forests. Lianas can completely cover a tree canopy, severely restricting access to light and adding weight that the tree must support. Lianas often link several tree canopies together, so when a tree falls, it may pull down others with it. This larger canopy opening provides more space for shade-intolerant trees, non-native invaders, and more lianas. Their rapid growth rates make lianas excellent colonizers, with some species persisting long after the canopy has reclosed.

## The Sweet Smell of Success

Japanese honeysuckle (*Lonicera japonica*) is a viny member of the same genus as the bush honeysuckles discussed previously. It is also the honeysuckle of my childhood. Their flowers produce lots of nectar, the honey of honeysuckle. As children, we would pick the flowers and suck the sweet nectar out of them as a summertime treat. There are several native honeysuckle lianas, predominately in the southeastern United States. These have beautiful flowers, yet I have sadly seen native representatives in the wild only a few times. They are much more common in gardens than in nature. The scent of Japanese honeysuckle perfumes humid, early summer nights with a sweetness that approaches overpowering. It was a sad day when I learned that this was not a native species.

Japanese honeysuckle was introduced into North America from its native eastern Asia for horticultural purposes precisely for the reasons noted above, the pretty white and yellow flowers and that wonderful scent. I found a historical record of this species when I worked at Willowwood Arboretum in New Jersey. In his expansive notecard system, Benjamin Blackburn, a meticulous note taker and caretaker of the property for decades, annotated his species record with "a most pestiferous plant, removed." That is about the most succinct yet accurate species description that I have seen.

The species is a twining vine that climbs by twisting itself around other plants rather than using coiled tendrils as does a grape, or adhesive projections, as with English ivy. The vines are small when young, twining up herbaceous plants such as goldenrods or woody species to reach up into the sunlight. If they colonize a tree, we often find several Japanese honeysuckle vines twining up each other in their race for the canopy. The species is smaller than many lianas, with stem diameters rarely exceeding 1 inch (2.54 cm). Japanese honeysuckle excels at covering shrubs and small trees in a carpet of their leaves and can pull established individuals down under their weight. At a minimum, they severely reduce growth in small trees. The smallish maximum stem diameter keeps Japanese honeysuckle from attaining dominance in taller canopy trees, restricting this species to early successional forests. It preferentially grows on two early successional trees, eastern red cedar (*Juniperus virginiana*) and flowering dogwood (*Cornus florida*, chapter 6), and avoids later-successional trees such as maples (*Acer* spp., chapter 5) and oaks (*Quercus* spp., chapter 6). This habitat selection likely occurs because the dense branching structure of the early successional trees allows the liana to twine its way across the trees' canopy.

With such prolific flowering, we might expect Japanese honeysuckle to spread excessively by seeds. The species is self-incompatible, requiring pollen transfer between genetic individuals to produce seeds, thus constraining prospective mating opportunities. The flowers are visited by bees, primarily concerned with pollen collection, and nocturnal hawkmoths (Sphingidae, the sphinx moths). Bees aren't great at pollinating the flowers, but the hawkmoths are excellent. Hawkmoths are to most moths the same way that hummingbirds are to most birds; they hover in place and feed at nectar-producing flowers. Because of their massive energy needs, they visit flower species with lots of nectar and are willing to travel long distances to find good sources. Long-distance flying makes them excellent vectors for gene flow. However, there aren't enough hawkmoths for Japanese honeysuckle, so fruit production is typically low—often only a few of the quarter-inch, black berries on each plant.

The lack of fruit and seed production limits the rate that this species can spread. However, the species easily roots along its stem, supporting more aboveground growth and allowing the species to become locally dominant. Many of the honeysuckle stems on a tree, maybe even on multiple trees, are genetically identical and, therefore, inappropriate mates. Japanese honeysuckle has two modes of reproduction, vegetative and sexual, that operate at different spatial scales. Vegetative expansion allows local spread and resource capture, whereas fruit production, with the assistance of fruit-eating birds, allows larger spatial spreads and can establish populations in new habitats.

Japanese honeysuckle covers herbaceous and woody species alike with dense foliage that persists late into the fall. In the absence of trees to support vertical growth, Japanese honeysuckle can form a solid carpet of foliage, smothering competing vegetation. When shrubs or trees are present, the liana can cover lower branches or completely engulf small trees and shrubs. Well-established trees may be tall enough to avoid much of this damage, but tree seedlings are particularly vulnerable. In addition to the tree's foliage being covered by the liana's, the twisting growth of Japanese honeysuckle can strangle a young tree, cutting off its vascular tissue. Larger trees, however, may also still be affected by honeysuckle as it can provide a climbing structure for larger lianas, such as the following species.

### Twisting into the Canopy

Asiatic bittersweet (*Celastrus orbiculatus*) is another exotic invasive with a native relative (*C. scandens*, American bittersweet) that is much less common. Both are beautiful plants grown in gardens for their winter fruit displays, which persist after the leaves have fallen. The fruit's orange outer layer splits open to reveal a bright red berry that birds disperse. These fruit displays are stunning in the winter's landscape and are often brought in as holiday decorations. It was precisely this characteristic that motivated the species' introduction from eastern Asia around 1860. Asiatic bittersweet spread rapidly after introduction, now occurring over most of the eastern United States and southern Canada. The native and the exotic are very similar—both are primarily dioecious (separate male and female plants), both grow by twining up trees, both grow well in forest gaps. So, why is the invasive Asiatic bittersweet so much more successful, with growing populations, whereas populations of the native American bittersweet are declining?

The invading species produces more offspring than does the native and has faster growth rates; thus it performs better than the native species. Moreover, Asiatic bittersweet has one additional trait that gives it a distinct advantage—it can forage for light. As light passes through plants, the light wavelengths absorbed by chlorophyll and other leaf pigments become depleted, particularly true of the red wavelengths. Red light is strongly absorbed, depleting it in the available light, while longer, far-red wavelengths remain unchanged. This change in the "color" of light, the ratio of red to far-red light, can be detected by many plants to inform them of their environmental conditions. Light low in red wavelengths relative to far-red indicates the presence of other plants above them.

The exotic Asiatic bittersweet can detect this color shift and alter its growth by increasing stem and leaf production allocation. Under full sun conditions, the exotic grows twice as fast as the native; under light depleted in red wavelengths, this difference increases to fifteen times the growth of the native. The change in growth patterns allows the plant to search out light actively while growing on the forest floor and climb into the canopy rapidly when given the opportunity. The invader also persists for much longer in shaded understories, awaiting an opportunity to enter the canopy. I have often seen the vines lying on the forest floor for several feet until they ascend a tree trunk. The ability to persist under shade and seek out better light conditions tip the competitive balance in favor of the invader.

As a twining vine, bittersweet coils around small trees like the previously discussed Japanese honeysuckle. However, honeysuckle is a small vine that exerts little pressure on a tree. Asiatic bittersweet is much stouter and can effectively strangle a tree as the tree and liana increase in diameter. Sapling-sized trees are often killed in this way. Once in the canopy, the lianas compete with their supporting trees for light, and the massive size of the lianas often snaps trees during storms. Asiatic bittersweet infestations can be so dense as to effectively eliminate tree regeneration, shifting the system away from a tree-dominated forest. Trees at forest edges or in hedgerows are particularly at risk as the lianas grow even more quickly in the higher light and can rapidly pull trees down under their weight.

The two bittersweet species are very closely related to each other. They are so closely related that they can form reproductively viable hybrids. The ability to hybridize is even worse for the native bittersweet. As the exotic Asiatic bittersweet increases in the landscape and the native American bittersweet decreases, the closest reproductive partner for the native will often be the exotic invader. Conservation of existing native populations will be hampered as the fruit produced may not always be purely the result of crosses within the species. Over time, it may become

difficult to separate the two species, and the native may become genetically lost.

## Herbs

The vast majority of non-native herbaceous plants are those of open,
sunny environments. Once a forest gap closes, most invasive herbs disappear quickly. For this reason, we will forego looking at early and late-
successional herbs here and instead focus on two forest invaders. There are
many native herbs in forests, but remarkably few non-native species are
more than occasional members of the forest floor, and fewer have sizeable
effects on the system. Native understory herbs are also typically perennials, taking years to accumulate the resources necessary for reproduction.
Annuals are typically rare in shaded environments as there are often insufficient resources to grow, flower, and produce seeds successfully in a single
growing season. The two herbaceous species that we will focus on are
both short lived, an annual and a biennial (a 2-year life cycle), in marked
contrast to the perennial life cycle of most competing natives. Both are
genuinely frightening invasions with disastrous implications for forests.

### Chemical Warfare under Our Feet

Garlic mustard (*Alliaria petiolata*) is native to Europe and was introduced
to North America as a culinary herb before 1868. As may be surmised by
the common name, the plant is in the mustard family, and it tastes and
smells like garlic. I have eaten it on more than one occasion and have
never found it a pleasant taste, so I question the reason for introduction.
This plant is a biennial, taking two growing seasons to grow from seed
before it reproduces and dies. In its first growing season, it germinates
and forms a rosette of heart-shaped vegetative leaves, typically growing
late into the summer and fall. From the resources accumulated during
this first period of growth, garlic mustard will bolt early in the spring of
the second year, flowering and producing seeds. The fruit is typical of all
mustards, splitting apart and spilling the seeds on the ground. Walking
through ripening garlic mustard in early summer causes the fruit to split
and scatter everywhere, recolonizing the space opened by the now-dead

parental generation. The seeds easily stick in mud, relying on the hooves of deer and other animals for site-to-site movements.

Besides not accumulating energetic reserves for multiple years, populations of garlic mustard must constantly rely on seeds to re-establish the population. This dependence on restarting from seed would seem to put the plant at an extreme disadvantage to nearly all perennials that may live for decades or more and hold onto the same piece of real estate growing season after growing season. Garlic mustard, however, has a sinister way of leveling the field with its native competitors—allelopathy. We briefly discussed allelopathy, plant–plant chemical warfare, with black walnut (chapter 6) and *Ailanthus* (this chapter) as a way to inhibit the performance of competing plant species. However, garlic mustard affects a fungal mutualist, indirectly affecting competing plants, shifting the ecological balance in garlic mustard's favor.

Mycorrhizae, literally "fungus roots," are mutualistic associations that fungi form with the roots of most land plants. Pro-fungi biologists argue that the only reason land plants evolved is because of such partnerships with fungi. Though individual plant–fungi combinations may be more or less beneficial, mycorrhizal associations benefit most plant species, improving growth. Because the fungal hyphae, the threadlike body of fungi, are so thin, they contact an immense soil volume, much larger than a plant root alone could. The fungus can digest (decompose) organic matter in the soil and provide the plant with greater access to water and mineral nutrients, particularly phosphorus. In exchange, the fungi extract sugars from the plant roots to use as a food source. As carbohydrates are often in ready supply, this is a cost that most plants willingly pay. Nutrients, simply put, are more limiting than energy in the form of carbohydrates to a plant.

Most land plants form mycorrhizae. Plants tend to produce most of their roots within the top few inches of the soil, where most resources are located. All plants exploit this same layer for water and mineral nutrients, from the smallest herb to the tallest tree. Most of these plants depend on mycorrhizae to assist in resource uptake. The chemical damage to the fungal community generated by garlic mustard affects other forest herbs within this soil layer and almost certainly affects canopy trees. Mustards, as a family, do not form mycorrhizae. They have too weedy of a life cycle (rapid growth and reproduction) to make putting energy into a long-term fungal relationship a beneficial prospect. Mustards naturally colonize open patches of soil that typically have freely available soil resources in them. With no competition, the plants can complete their life cycle before other plants recolonize. Garlic mustard releases allelopathic chemicals that act against mycorrhizal fungi, reducing the number of fungi associated with other plant's roots. Removing the partnering fungi subsequently removes the benefit of enhanced resource acquisition for any plant dependent on mycorrhizae. As garlic mustard does not form mycorrhizae, its root function continues unabated.

As a garlic mustard invasion persists, the density of mycorrhizal fungi in the soil decreases. More importantly, associated understory herbs' nutrient uptake is hampered as the fungi decline. Storage in belowground structures may buffer other plants for a time, allowing continued success. Eventually, growth and reproduction decrease enough that native plant populations decline, allowing more space and resources for garlic mustard. Areas with heavy invasions often have reduced native herb cover and diversity, driven by disruption of the mutualistic fungal relationship upon which the natives depend.

Degradation of forest understory flora is how we tend to perceive the impacts of garlic mustard. Primarily we do this because the invader occupies this forest stratum, and we see the plants next to each other, interacting. We often think competition with canopy trees is severely one-sided in favor of the trees, but this is one situation in which an understory herb may significantly affect tree growth.

## An Improbable Plant

Japanese stiltgrass (*Microstegium vimineum*), a native of southern and eastern Asia, first appeared in Tennessee in 1919, allegedly transported as packing material for a shipment of porcelain. From this initial location, it has spread outward to all eastern states except Maine and has moved into Ontario, Canada. It is a forest understory grass that tends to form vast, single-species swards in forests. It does best in wetter areas, but drier habitats and sometimes full-sun habitats also appear suitable. While forests containing Japanese stiltgrass appear lush and green at first glance, they lack the richness of species that make understories the reservoir of forest diversity. It is an odd species with an unlikely suite of characteristics that make it a great invader despite itself.

The improbability of Japanese stiltgrass becoming invasive comes from the way it does photosynthesis. Most plants use a photosynthetic pathway that results in a three-carbon sugar as the first stable product. A six-carbon sugar briefly forms but immediately splits into two halves. This baseline pathway is therefore known as $C_3$ photosynthesis. This metabolic process works in many situations and is the most common way that land plants earn their livelihoods. However, problems appear under hot and dry conditions when the plant tries to conserve water by closing down its stomata (leaf pores). This action conserves water but makes internal carbon dioxide levels drop, triggering the photosynthetic machinery to become metabolically destructive. Some plants that often find themselves in environmental conditions that generate this destructive biochemical behavior have evolved an additional step in the process. These species use an additional enzyme to concentrate carbon dioxide internally, forcing the existing $C_3$ photosynthetic machinery to operate appropriately. They produce a four-carbon sugar as the first stable product, known as $C_4$ photosynthesis. The critical feature here is that the $C_4$ pathway retains the entire $C_3$ process, just with the addition of a few enzymes and many carrier molecules.

Because of the costs incurred by these additional metabolic pieces, $C_4$ species occur in full sun and drier habitats, where the benefits of the pathway are greatest. Shady, cool, and wet environments remain the domain of $C_3$ plants. Photosynthesis via the $C_4$ pathway has evolved several times within the grasses, and this is the plant family that contains most of this pathway's occurrences. The $C_3$ grasses are more abundant northward, replaced by $C_4$ grasses as you move south. A simple example from agriculture is that wheat ($C_3$) is a northern, cool-season crop and corn ($C_4$) is a warm-season crop of central and southern North America. Based on this admittedly long explanation, we would expect Japanese stiltgrass, a plant of moist forest understories, to be a $C_3$ grass, but it is not. As a $C_4$ grass, we would expect the costs of the additional photosynthetic machinery to be insurmountable in the shade, but somehow it works.

We may envision a scenario in which a plant gradually accumulates energy year after year, growing slowly and eventually reproducing. Such a strategy could be successful under the right circumstances, particularly when year-to-year survival was high. Unexpectedly then, Japanese stiltgrass is an annual. There is no accumulation of resources over multiple growing seasons that make up for an inefficient photosynthetic process. The only resources held onto are the energy (carbon) and nutrients that provision each seed. And the seeds are small, a competitive disadvantage. Quite simply, physiology tells us that this species should not be invasive, let alone a viable forest species.

If we look beyond physiology, the invasiveness of Japanese stiltgrass begins to make more sense. As an annual, the species produces many seeds, the only way to contribute to future generations. Uncertainty can be deadly for a species dependent on seeds each year, so Japanese stiltgrass employs a bet-hedging reproductive strategy. As the plants grow, they produce tiny flower clusters entirely enclosed in a leaf, never exposed to the pollinating wind. These flowers self-pollinate and represent an insurance policy against any late-season droughts or other challenges to reproduction. If the plant lives to the end of the growing season, it will produce exposed flowers at the tip of each stem to be pollinated by the wind, hopefully from another individual. These two types of flowers ensure at least some reproduction in all years, with the addition of some outcrossed seeds in good years to provide new genetic combinations.

Since the seeds are tiny, they rely on exposed soil to germinate and become established. There simply is not enough energy within a seed to push past any leaf litter in the forest understory. White-tailed deer come to the rescue here. When deer move from forest patch to forest patch in the winter, looking for food, they disperse the seeds as they forage. In areas with high densities, deer often trample the leaf litter, exposing bare soil and generating places for stiltgrass to germinate. Deer do not eat Japanese stiltgrass, focusing instead on other forest herbs. This selective feeding removes potential competitors, further allowing expansion of the invader. The grasses are weak stemmed, sprawling over the forest floor and other plants, reducing access to light. In the winter, stiltgrass litter decomposes very quickly, providing bare soil for the next generation of seeds.

In my research, I have watched this species sweep across forested areas, increasing from an occasional occurrence to dominance in five or so years. Deer had severely affected some areas, so there were few plants in the understory—an open and available ecological niche. However, other areas had well-developed herbaceous understories when the invasion occurred; within a few years, diversity decreased dramatically, as had the total abundance of other species. This firsthand experience, more than any other, has clarified the ecological horror that invasive non-native plants represent for me. The species is in Indiana, just east of my current location, and just south in southern Illinois. I know it is only a matter of time before it arrives here, but I still obsessively clean my boots so that I am not the one to transport it here.

8

# ACCUMULATING IMPACTS —
# PUTTING IT ALL TOGETHER

What I have attempted to do in the preceding chapters is to present a species-by-species account of the threats to the forests of Eastern North America. While this approach has allowed me to feature the unique ecological attributes of the trees and their threats, I fear it may lead to a disconnected view of the challenges facing our forests. Dealing with each tree or invader separately may inaccurately portray the threats as separate and not interconnected. The harsh reality is that many, if not most, forests are facing multiple threats. All Eastern forests face issues associated with climate change, exotic plant invasions, and likely at least one tree species loss. A forest facing only one issue is a lucky forest indeed.

All these challenges can be thought of as issues of forest health. This admittedly vague term is helpful as it possesses an analog in human terms. From a human perspective, we may view health as feeling good, with all our systems functioning reasonably well. We face acute incidents, such as broken bones and infections, and chronic conditions, such as arthritis and asthma, as we go through life. We also have conditions that can predispose us to other health issues, such as high blood pressure and kidney failure or diabetes and heart attacks. Our bodies' systems are integrated, and so our health issues are as well. One medical issue may not mean much to our health, but accumulate several issues, and health declines or our grasp on health becomes more tenuous, more easily disrupted to disastrous results. We may not have a good definition of health, but we know when our health is failing!

We can subdivide forest health issues into similar categories to those of people. Acute incidents may include a storm that generates gaps by knocking over trees or a fire that sweeps through the forest understory and consumes the accumulated leaf litter. In moderation, these incidents can be beneficial to a forest's overall health. Chronic issues would include climate change, acid rain, and long-term increases in deer populations. These chronic issues may represent pre-existing conditions that make invasions by non-native plants or non-native forest pests more likely to happen. For example, plant invasions are strongly linked with human disturbance of forests, namely logging. The disturbance itself does not cause the invasion; rather, the disturbance allows an invasion to happen if a non-native shrub or tree is close enough that its seeds can colonize the newly opened habitat. As issues accumulate, the forest becomes more unhealthy—its functioning as a habitat and natural resource becomes hampered. It would still be a forest but with diminished ecological, emotional, and economic value. We would recognize this as a crappy forest, in colloquial terms.

My goal in this chapter is to briefly describe the history of one forest to show the compounding impacts of multiple changes within its boundaries. While this is not a unique forest, it serves as a cautionary tale for all forests. It has faced monumental changes over time and will without doubt face more in time. The forest that we see today results from the forest's history and the accumulated challenges to forest health that it has faced.

## Mettler's Woods—a Sylvographical Study

For some people, biographies are interesting reads—they place you in the shoes of people in whom you are interested, hopefully in a compelling way. Famous people, or people with otherwise fascinating stories, are common subjects because we want to know how they got to where they ended up, for good or ill. Perhaps we can pattern our own lives after theirs or learn the pitfalls to avoid without suffering through similar painful experiences. Like anyone's story, biographies contain trials and triumphs, uncertainties and dramatic occurrences, births and deaths. Forests, sylvan habitats, also have their own stories, their sylvographies, if you will. Perhaps a sylvography will not naturally be as compelling as the life story of a celebrity or historical figure. However, the stories of forests will still contain drama and perhaps a lesson or two to be learned.

I want to present the sylvography of Mettler's Woods, the raison d'être for Hutcheson Memorial Forest, a property of Rutgers University where I did my dissertation research. It is an old-growth forest that has never been logged, but that does not mean that it has not seen its share of trials and tribulations. The story of this forest is like those of many forests across Eastern North America, just better documented than most because of its university association. At a certain level, this forest has just become more famous than others, a sylvan celebrity of sorts.

As with all forests, Mettler's Woods has had a long history of interactions with people. Before Europeans "discovered" and then settled the east coast of North America, native peoples such as the Lenni Lenape, also called the Delaware people, managed the forest with fire. Frequent fires cleared out the forest undergrowth, pushing the forest composition toward fire-tolerant trees that also conveniently increased the abundance of wildlife-feeding trees such as oaks (*Quercus* spp.) and hickories (*Carya* spp.). Fires occurred in these forests every 10–25 years, as documented by fire scars on the oldest trees. These were primarily low-intensity fires that rarely killed established trees, but the fires were strong and consistent enough over time to shape forest composition and structure.

The Dutch were the first European people to settle the area, establishing New Amsterdam in 1624 and building farms throughout the area. The Mettlers, the name the forest now bears, were one of these farming families. As early settlers arrived, the burning stopped, as no fire scars occur after 1711. The cessation of fires changed the forest, allowing tree species that do not tolerate fires to expand in abundance over time. The absence of fire is likely changing the forest today—a remnant of a centuries-old impact. The changes are not immediately noticeable because most forest species are long lived and slow growing, taking decades to respond. Sugar maples (*Acer saccharum*, chapter 5) and American beech (*Fagus grandifolia*, chapter 6) are pretty intolerant of fire, and old trees of these species are restricted to a few patches within this forest. These two species have expanded somewhat in the site during the intervening centuries, but this expansion has been slow—established trees need to die before there is an opportunity for a new tree to reach the canopy. If the forest were left alone, the complete transition to a non-fire-dominated forest type may take additional centuries.

Over the ensuing decades after initial settlement, local farms expanded, roads were built, New Amsterdam was renamed New York, and one of the most densely human-populated areas on the planet began to affect the native vegetation. Most of the surrounding forests were cleared at least once, if not multiple times. Farms became more numerous for a while, ultimately to be displaced by housing developments, both land uses affecting the environment. For whatever reason, Mettler's Woods escaped the ax, then the chainsaw, and then became noted locally as a forest gem worthy of study. In 1954, a particularly active hurricane season (Carol, Edna, and then Hazel) blew down many large old trees, initiating talk of a salvage logging operation. Word of the impending doom of the forest got out, concerned citizens united, and the Hutcheson Memorial Forest Center was created in 1955 to protect the old-growth forest for perpetuity. Unfortunately, the forest is protected only from active damage, in other words, logging. It is yet to be wholly protected from the actions of people or from natural threats like hurricanes and climate change.

The forest is poorly drained with shallow soils, mainly 3 feet (1 m) or less over shale bedrock. The soils often become saturated with water in the winter and spring when evaporation is low. For this reason, this location never had a population of American chestnut (*Castanea dentata*, chapter 2), though much of New Jersey initially did. Chestnuts require a reasonable amount of soil moisture but cannot grow in poorly drained soils. Instead, this forest had much American elm (*Ulmus americana*, chapter 1), a species tolerant of seasonally waterlogged soils. The American elm canopy largely disappeared from this forest when Dutch Elm disease erupted in the 1930s, though a few small trees persist to this day, yielding seedlings from time to time. They never last long, so the American elm is functionally lost from this forest.

In the 1980s Mettler's Woods still retained a nearly continuous understory of flowering dogwood (*Cornus florida*) over 20 feet (6 m) high. The historical photos of these trees crowned with their large, white blooms are spectacular! These trees regenerate in more open conditions, so they were likely remnant reminders of the open, fire-dominated history of the forest. In the 1990s, a fungal anthracnose disease swept through the population, killing many larger dogwood individuals within the forest (chapter 6). Few of these trees remain in the old-growth forest, though they are abundant in the abandoned agricultural fields, which are now young forests surrounding the old-growth stand. Some of these newer forests have now developed dense enough canopies that the dogwood anthracnose has re-appeared.

In the early 1980s, invasive spongy moths (*Lymantria dispar*; chapter 6) made it to the site, devouring their favored forage species—the oaks. The moth larvae defoliated the oak portion of the forest canopy to the extent that there was full sun reaching much of the forest floor in the middle of the summer when a forest should be at its shadiest. Oaks can tolerate that level of defoliation once, but the outbreak of spongy moths continued a second year, killing some of the ancient oaks. In this new, high-light environment, shade-intolerant tree species such as ash (*Fraxinus* spp.) became established, leading to a nearly even-aged cohort of trees that persists today in the areas where the spongy moth damage to the oaks was the greatest. These new forest colonists, the outcome of a species invasion themselves, are now threatened by the emerald ash borer (*Agrilus planipennis*; chapter 4), which arrived in the forest around 2015. These tree species will likely disappear from the Mettler's forest as well. Transitioning away from ash would follow the regular compositional change from shade-intolerant to shade-tolerant tree species. However, future gaps that occur in the forest will be missing their key colonists, the ash, leaving room for something else to colonize.

Over its long history, Mettler's Woods has also seen many new plant species arrive from around the globe. First came the early settlers' accidental transmission of agricultural weeds that had no chance of colonizing a mature forest. Many of these plant species persist as problematic weeds of open habitats today, such as clovers (*Trifolium* spp.), ox-eye daisies (*Leucanthemum vulgare*), and Queen Anne's lace (*Daucus carota*, also known as wild carrot). Later, species more adapted to shady forest conditions were introduced, intentionally and otherwise. People are fond of gardening, and many of the same characteristics that make plants attractive to gardeners—pretty fruit, good growth, flexibility in their growing conditions—are also characteristics of an invasive species. The increasingly suburbanized landscape that surrounds Mettler's Woods serves as a seed source for these invaders. Of particular note in the old-growth forest are two trees, Norway maple (*Acer platanoides*, chapter 7) and tree of heaven (*Ailanthus altissima*, chapter 7). Norway maple is now fully established and reproductive in the old-growth forest, generating many smaller trees poised to make their way into the canopy as opportunities allow. *Ailanthus* now functions as a classic gap-colonizing species in Mettler's Woods.

The forest understory has also changed dramatically over time. The East Coast has been experiencing an overpopulation of white-tailed deer (*Odocoileus virginicus*) for decades, affecting the forest. When I first arrived on site in the mid-1990s, I often walked through the old-growth forest to my research site located toward the back of the property. In the spring, I would walk through spring beauties (*Claytonia virginica*), mayapple (*Podophyllum peltatum*), trilliums (*Trillium sessile*), and jack-in-the-pulpits (*Arisaema triphyllum*) regularly. As deer populations increased, many of the spring wildflower species declined, along with the density of native tree seedlings and saplings. The open forest understory left behind was invaded by Japanese stiltgrass (*Microstegium vimineum*, chapter 7), which quickly came to dominate. I first noticed this species in the old-growth forest somewhere around 1995. I knew the species only because other students had been working on its invasion elsewhere. Japanese stiltgrass now forms a nearly continuous green carpet in the old-growth forest, except for under a patch of sugar maples that cast too much shade for the grass to persist. This small area has resisted the invasion for more than two decades now. Over 12 or so years, I watched Japanese stiltgrass take over the forest floor and displace much of the remaining understory flora.

I also remember walking along that path in an area where spicebush (*Lindera benzoin*) arched overhead. It is a lovely native shrub with an even lovelier scent when you crush its leaves between your fingers, almost certainly a chemical defense. This shrub is a larval host for the spicebush swallowtail butterfly (*Papilio troilus*) and produces lipid-rich fruit consumed by migrating bird species. The last individuals of the spicebush patch died during a late-summer drought in the 1990s. Many other shrubs are also largely missing from the forest now. Researchers in the 1950s were able to differentiate the forest plant communities of Mettler's Woods by the shrubs present in the forest understory. This ability seems a foreign concept now, with the shrub species largely eliminated or replaced by invasive exotic species. One community at the site was defined by the dominance of maple-leaved viburnum (*Viburnum acerifolium*). I have seen that shrub in the forest only as a single seedling and one small adult. Another native viburnum, arrowwood (*V. dentatum*), used to be nearly everywhere but is now restricted to a few small patches along a stream. The loss of shrubs is likely a combination of a more closed forest canopy resulting from the absence of fire, increased browsing by deer, and new competition from more aggressive non-native plant species.

While the native shrubs of Mettler's Woods have declined dramatically, the non-native component is thriving. The forest has well-established populations of several shrubs used in home landscaping. These include common privet (*Ligustrum vulgare*), Japanese barberry (*Berberis thunbergii*), and bush honeysuckles (*Lonicera maackii* and *L. tatarica*, chapter 7) in shaded areas. Autumn olive (*Elaeagnus umbellata*) and *Rosa multiflora* occur along edges and in larger gaps. Wineberry (*Rubus phoenicolasius*, chapter 7) appeared in the old-growth edges around 2000 and has since filled many forest gaps. The most recent addition to the forest is the Japanese angelica tree (*Aralia elata*), a super-spiny small tree or large shrub that spreads quickly through bird-dispersed fruit. To this list, we can add a variety of non-native lianas. English ivy (*Hedera helix*) is present but at meager numbers. Japanese honeysuckle (*Lonicera japonica*, chapter 7) is now commonly found along edges and in large gaps. Asiatic bittersweet (*Celastrus orbiculatus*, chapter 7) is increasing dramatically, sometimes using Japanese honeysuckle to accelerate its climb into the canopy. This liana has pulled down some trees adjacent to the old growth stand, and it is likely to continue spreading and taking down trees in its wake.

In an attempt to curtail the effects of white-tailed deer, land managers of Mettler's Woods installed a 10-foot-tall fence around the old-growth forest in 2015. A fence this tall is enough to discourage but not completely exclude deer. They are cunning at going over or under the fence, and the forest itself is constantly opening gaps in protection, allowing deer access. Local archers now hunt the area and donate the culled animals to food pantries. Despite the high burdens of non-native shrubs and herbs already present, some recovery has occurred. Seedlings of jack-in-the-pulpit and arrowwood have appeared for the first time in decades, and there has been an overall increase in the number and variety of native tree seedlings. Whether deer management will curtail any non-native plant species is unknown yet, but it is certainly possible.

The most recent invasion into Mettler's Woods was not another plant but an insect pest. The spotted lanternfly (*Lycorma delicatula,* chapter 7) appeared in 2019 and has since gone through a population explosion. This invasive insect favors tree of heaven and native grapes (*Vitis* spp.) in the forest. As sap-sucking insects, lanternflies allow much of the sugar-rich plant sap to pass through their digestive system so that they can harvest the low-concentration protein. This sugary liquid rains down on the vegetation underneath, causing mold to grow on understory plants. Whether this is just unsightly, will interfere with photosynthesis, or will increase bacterial and fungal infections is not clear. However, the insects are pulling a lot of energy from their hosts, yielding almost certain decreases in plant performance.

There is also an accumulating list of chronic changes to the conditions of Mettler's Woods. The forest is now wetter and warmer than it has ever been. As both moisture and temperature (as well as their variation) are essential in determining critical biological rates such as photosynthesis and decomposition, changes may have been accumulating for decades. Nitrogen, the core component of amino acids that form important proteins such as enzymes, is typically the single most important limiting nutrient in terrestrial systems like Mettler's Woods. Our continued dependence on fossil fuels results in nitrogen deposited with precipitation at rates sufficient to alter the success of plant species. Increased nutrient availability may ultimately favor species that are less conservative in their nutrient use strategies, such as many of the non-native invaders.

Herbivorous insects also tend to be limited by nitrogen, feeding on more plant tissues than their caloric demands require so that they may acquire enough nitrogen. Greater nitrogen availability may translate to larger densities of herbivorous insects. The ratio of carbon to nitrogen in plant tissues is also essential for decomposition. Higher nitrogen availability and warmer temperatures should lead to faster decomposition rates, with untold effects on soil organisms. All these changes to the system may cascade throughout the forest, altering in subtle ways a whole host of processes.

The problematic issue is that we likely won't know the implications of these accumulated changes to Mettler's Woods without careful experimentation. Scientists and society appreciate simple, straightforward, cause-and-effect stories. Complex interaction webs are difficult to unravel and to understand. We may never truly understand the subtle web of these contingent processes for this or any of our forests. Perhaps that is not a particularly satisfying way to end a sylvography, but the story of Mettler's Woods is not yet complete; I hope it never is.

## How Typical Is Mettler's Woods?

What makes biographies compelling is the height of the challenges faced and the depths of the trials confronted. Without these, the story could be that of any ordinary citizen. We rely on the drama faced to form a compelling story that resonates with us. The story of Mettler's Woods is an excellent example in that vein. It illustrates the worst that can happen to a forest outside of logging, a hurricane, or other catastrophic events. But likewise, applying the narrative of Mettler's Woods to all forests of Eastern North America would be horridly misleading.

Mettler's Woods occurs in one of the most densely populated human corridors on the planet and subsequently has received much more human impact for longer periods than most North American forests. As such, the level of invasion and disturbance and their impact are much greater. The forest stories outside this immediate area are often much less dramatic and much less depressing and, in many cases, show some real improvement over time despite their challenges.

The forests of many areas have regrown following early clearing for agriculture or timber production. More recent harvesting is typically done more responsibly now that the lessons of early logging have been learned. Leaving seed trees for natural regeneration, providing shelter wood to encourage shade-tolerant species growth, replanting to supplement regeneration, and removing exotic plants to minimize competition are much more common, though not universal. Similarly, at least minimal management occurs in many forests, whether it is implementing a fire regime or removing suppressed trees to encourage canopy tree growth. Healthier forests may still face one or more biotic challenges but have a much better long-term prognosis.

The key takeaway here is not to despair and give up. It is simply to keep horrid, worst-case examples such as Mettler's Woods in our minds as lines not to cross. A lesson learned and appreciated. If we value forests, and a great many North Americans do, they are worth conserving. The largest challenge for local conservation efforts is awareness of the issues that forests face. Exotic species removal, restoration of forest understory communities, and targeted planting when regeneration fails are nearly universal conservation approaches that can be useful in any forest. The themes of the next chapter deal with forests at regional scales and involve public policy and governmental regulations. Long-term, these may make local actions less necessary, but this is a long way off.

# CONCLUSION

## *Protecting Our Forests' Future*

After eight chapters outlining how bad things have gone awry in Eastern North America's forests, it can be easy to lose hope and throw one's hands into the air in dismay. In all honesty, I do not think this is a hopeless situation, undoubtedly dire, but something that we can address if we choose. Throughout this text I have outlined work researchers are conducting to save some tree species, and that research must continue. Genetic engineering and traditional plant breeding to increase tree resistance to diseases, together with biological control of pest insects, hold great promise in these areas. The key to long-term success in protecting our forests' future is breaking the cycle of invasion, loss, and remediation. This cycle is a massively expensive and reactionary way to address our forests. It is also an approach that has us continually in a state of recovery.

The adage that an ounce of prevention is worth a pound of cure is appropriate to our forest's dilemma. Prevention of new pests incurs costs, high costs. However, remediation is much more expensive. Insurance companies now recognize that it is far more fiscally responsible to pay for preventative care than to pay for treatment of a life-threatening illness later. Insurance companies do this because they pay regardless of whether the care is preventative or reactionary. In forest protection, prevention costs may fall on whomever imports forest products or shipping materials, cutting into profits or being passed on to the consumer. However, the costs of remediation typically fall on the government, dispersed across all citizens in the form of taxes. The cost of emerald ash borer (*Agrilus planipennis*, chapter 4) is estimated to be US$10.7 billion, and this certainly underestimates the continued loss of an important timber species. The entire 2021 budget of APHIS (Animal and Plant Health Inspection Service) was a little over $2 billion, with plant protection accounting for only a fraction of that. Perhaps greater inspection and monitoring efforts could have prevented the invasion or allowed eradication before the ash borer escaped, saving much of the costs we now face. Of course, we would never know how much money we saved, just the costs associated with unsuccessful efforts. Continued vigilance is costly and necessary, but most of us carry insurance policies, so we are familiar with the concept.

Here I briefly outline four strategies that could easily prevent or minimize future costs surrounding species losses and forest pests. As you will see, these strategies are not entirely independent from one another, but they are necessary to ensure healthy and sustained forests. These recommendations apply anywhere in the world, though they are presented here from a North American perspective. Again, forest issues are occurring everywhere, so this is truly a global issue.

## Strategy 1: Stop the Introduction of New Forest Pests and Pathogens

At first glance, this approach seems obvious and straightforward—find out how fungi and insects are arriving and block that pathway. It has always been evident that the entry path lies with international shipping. With the source so clearly known, what is preventing us from closing this invasion route? Quite simply, it is a question of economics. We like things to be inexpensive, and one way to meet demand and maintain low costs is to import goods. Clothing, electronics, and similar goods pose little threat, as their packaging commonly involves heavily processed paper products incapable of transmitting anything living into a new area. However, machinery, heavy tool parts, and other large items are often shipped using wood packing materials. Nearly everything is placed on top of a pallet for loading and unloading into containers. This wood appears to be the vector of nearly all modern forest pest transmission.

Removing or substantially reducing the risk of insect or pathogen transmission is simple, but it requires time and effort. Bark beetles and most pathogens live directly under the bark, so the wood used for international shipping should always have all bark removed. Bark contributes no strength to wood, so its removal does not affect structural integrity. The remaining wood will pose a biological threat only if any pests it harbors retain moisture and viability. Fungi and insects often have resistance to desiccation, making air-drying much less effective at limiting pests, but it is better than nothing. Wood products can also be kiln dried (drying facilitated by heating), heat treated, microwaved, fumigated, or otherwise chemically treated. All are effective at dramatically reducing the potential for pest transmission. However, it can be challenging to determine whether any wood packaging material has been appropriately treated once it is in transit.

The only way to entirely avoid issues associated with wood packaging materials is to avoid using timber for this purpose. Engineered wood products, such as oriented strand board, which is constructed from small wafers of wood glued together with resins, produce durable shipping pallets that are incapable of harboring pest species. The resins used to construct engineered wood products prevent insect or fungal pest growth. Even better, pallets made from recycled plastic should last even longer, cannot transmit forest pests, and generate no harvest pressure on local forests. There are likely many situations in which cardboard structures could replace wood as a support for heavy parts. I have seen some quite ingenious uses of corrugated cardboard in the last few years and am continually amazed by its flexibility.

Rules governing the use of wood packaging materials have been in place for years now, with some success. Any of the technologies described above fit the requirements for wood treatment. However, potential insect pests continue to arrive in North America. Many get intercepted at entry points, but the recent and repeated invasion of Asian longhorned beetles (chapter 5) tells us this effort has not been entirely successful. We also have greater regulation and surveillance of shipments coming from problematic areas, in particular eastern Asia. We need increased and continued vigilance to protect our forests, but this must happen in an economically responsible manner that minimizes impacts on shipping for those following the rules; those who do not follow the rules should perhaps financially support the system.

I propose an utterly simplistic inspection scheme, not entirely unlike how food safety inspections work. In my view, there are three inspection pathways, and the inspection fees should cover the costs of enforcement and serve as incentives to follow the rules. Low-risk shipments would be those with minimal or even no wood shipping materials. Inspection of these shipments would be at normal levels, although penalties for miscategorizing a shipment should be high enough to deter cheats to the system. Electronics, toys, and clothing should easily achieve this level of inspection and have minimal delays. Higher-risk shipments, those containing wood shipping materials or shipping wood itself, should have higher inspection rates and pay more for this requirement. Inspectors would determine whether the shipping materials meet requirements and examine them for any pest organisms. Any shipment failing this inspection would incur substantial fees to cover the actual costs of quarantining, destroying shipping materials, and any other necessary measures. Here enters the third inspection pathway, the naughty list. If a shipment is in violation, all further shipments from that group, including those without wood shipping materials, would be inspected for compliance, with fees set to cover those increased costs. There is a large import market for wood products; these need to be inspected heavily, but it may be as simple as a moisture test (probes do this quickly) to confirm that the wood is kiln dried.

Stopping the flow of new pests into North America is the foremost priority to protecting forests. If we can effectively do this, the following approaches become much less critical.

## Strategy 2: Prevent New Invasive Plants, Manage Those We Have

There are an estimated 5000 non-native plant species in North America. Agricultural activities accidentally introduced many of these soon after

European colonization. Today many invaders are intentionally introduced for horticultural purposes. The shift of plant movements away from agricultural weeds means that the species introduced are no longer restricted to the live fast–die young strategies that promote success in continually disturbed systems such as agricultural lands. These weedy species do not typically pose a threat to forested systems or only temporarily do so. However, horticulturalists select species to cover the wide range of environmental conditions that may occur in home gardens. A popular gardening mantra is "the right plant for the right place," reflecting the need to match a species' requirements to the site. An outcome of this approach is the need for a broad palette of plant species to suit any gardening situation.

Introducing plant species for shade gardens is particularly problematic for forested systems where these species may also thrive. Because we like interesting plants in our gardens, we typically appreciate flowers and fruit, which are so much more interesting than acquiring another shade of green. Horticultural varieties often invest a lot of resources into flowers and, if successfully pollinated, fruit. We particularly like brightly colored fruit for fall and winter interest. This trait also tightly matches bird dispersal of fruit, which may spread seeds across a landscape even if your backyard is not close to a forest. Many invaders build up populations in towns before making the leap to natural areas.

Growers like plants that are easy to propagate, either vegetatively or by seeds. This selection favors plants that can grow rapidly under the proper conditions, such as the higher nutrient conditions of a nursery. These plants, when they escape cultivation, tend to be more opportunistic if provided with adequate light, as found in a disturbed forest or at a forest edge. Shrubs and woody vines are particularly problematic in forested systems as they often reduce tree regeneration, leading to the long-term decline of a forest if the invaders persist. However, herbaceous species that become dense can also limit forest regeneration and affect the forest understory.

Although plant species from any temperate zone can invade forests in Eastern North America, most of the worst invaders have come from eastern Asia. How do they continue to arrive? There is still a thriving interest in acquiring something new for one's garden, supporting an industry devoted to new plant introductions. We tend to think of plant explorers as a thing of the distant past, but they still exist. It is easy to import plants as either cleaned seeds (no fruit allowed) or as plants as long as they can pass inspection—another APHIS job. Plant collectors continue to explore Europe, Asia, and, infrequently, temperate South America, searching for new species to introduce to a public willing to pay. Growers today can patent their plant introductions, protecting their investment in the collection and development of new varieties.

I can provide one example of how easy importing a potentially invasive plant is from my brief time working at an arboretum. There was, and still is, a seed-swapping group called *Index seminum*. They publish lists of species that locals collect either within their gardens or from the wild. There are members of the group from botanical gardens across the globe. If you are a member willing to share seeds from your location and promise not to exploit these plants commercially, you can request seeds. As the members speak many languages, the lists are typically Latin names numbered for easy order. Just select the numbers you want, and the seeds will arrive in the mail. As the arboretum curator, I received one of these seed packets, logged it into our system, and noted that one plant requested was replaced by a central Asian clover (*Trifolium*) species—just one number off from the actual species requested. Many exotic clover species have become naturalized in North America; the most common is white clover (*T. repens*) and is likely in your lawn at this moment. The potentially invasive clover I received in the mail could have easily been released.

Many countries, such as New Zealand, have much more restrictive species importing laws than the United States. These rules are designed to protect the country from unwanted invaders, and, as an island system, New Zealand has had a lot. Unless a species is expressly prohibited, and if the material passes phytosanitary inspection, there are no barriers to importing plants into either Canada or the United States. Problematically, prohibiting species is a political process, which does not function well or quickly. Legislators are hesitant to restrict any industry unless they are forced. Ecologists have generated many ways to assess species' potential to become invaders, but regulations do nothing without direct knowledge. The rules governing invasive plant species are typically not proactive, which is the only way to prevent an invasion before it begins. Although I am sympathetic to gardeners' desires for the new and exotic, I think it is well past time to stop allowing minimally regulated plant introductions. We have strong rules governing the movement of agricultural species; why can we not protect another agricultural resource, our forests, with a similar approach?

Even if all plant importation were to cease tomorrow, we would still have a massive burden of species already invading our forests or perhaps on the cusp of becoming invasive. We have lists of noxious species, plants that cannot be sold within individual states or provinces and, in some cases, are required to be controlled. However, inclusion as a noxious weed is no guarantee that the plant will not still be sold. Plants considered noxious in one state may be sold in an adjacent state, on the internet, or by an in-state vendor ignorant of or disregarding the law. Some states regulate many species, some few, some do not have any regulated species. With such a piecemeal effort, exotic plant invasions continue with little regulatory resistance. It is comforting that some states add species to their regulations before they are present as a preemptive measure, but this is not commonplace.

Some weed laws present pathways to force landowners to control species designated as noxious as a way to minimize their impacts. Again, with the sporadic nature of regulated species, some problematic species are not regulated, and enforcement of these laws is rare. In my current home state of Illinois, there is a list of noxious weeds that the government could enforce a landowner to control. *Cannabis sativa* is on the list for reasons other than being a weed (being weed, I suppose). There are three thistle species, musk thistle (*Carduus nutans*), Canada thistle (*Cirsium arvense*), and sowthistle (*Sonchus arvensis*; arvensis being Latin for "in the field"), and perennial *Sorghum* species on the list for their weediness in pastures. Two native ragweed species (*Artemisia artemisiifolia* and *A. trifida*) are famed for the allergic reactions they cause and are considered noxious in towns and incorporated areas. Only one species, kudzu (*Pueraria lobata*), is a weed of forested habitats. The state could make you remove populations of any of these species, but I have never heard of this being enforced.

Plants designated in the Illinois Exotic Weed Act are prohibited from sale or distribution in the state, but there is no mandate to remove established populations. This list is where most of the forest invaders appear. Regulated species include the following forest invaders: honeysuckles (*Lonicera* spp., chapter 7), multiflora rose (*Rosa multiflora*, Introduction), buckthorns (exotic *Rhamnus* spp.), kudzu, olives (*Elaeagnus* spp., not the edible kind), and Asiatic bittersweet (*Celastrus orbiculatus*, chapter 7). As these species are widespread throughout the state and region, regulating spread is helpful but is much too late at this point without active control.

Given the current gardening trend to select native species, many North American species could easily be cultivated and bred for garden usage, probably with fewer issues than non-native plants. Exploiting the plant diversity contained within this continent should easily offset the lack of new species from abroad. We need education, regulation, and enforcement to keep the exotic plants already invasive here manageable. We also need new regulations to support these efforts and monitor forested landscapes for invaders so that populations can be located and controlled before they become too abundant and control costs further skyrocket.

## Strategy 3: Minimize Pressure on Our Forests

Any forest harvesting activities generate disturbances in forests, whether from the tree removal itself or the damage associated with transporting the logs out of the area. There is no way to remove all disturbance from tree harvesting. Logging that uses helicopters or mule teams is less disruptive but slower and costlier and still causes damage by removing canopy trees. Selective logging, whereby individual trees or species are removed, leaving others undisturbed, is less damaging, but again, it still generates forest openings, roads, and tree drag paths. The less we disturb our forests, the less management they will require, and the healthier they will be. Minimizing forest impacts means minimizing our need for wood products. We cannot eliminate tree harvesting, but many things can reduce our dependence on new timber. If less wood is required, we can allow forests to mature more between harvests and likely improve wood quality and other species' habitat.

Within North America, wood is often a fuel source for at least supplementarily heating our homes. Typically, harvesting trees for fuel is low impact and selective. Fuelwood harvesting can easily be integrated into a forest management strategy. Trees that are crowded, not a desired species, or damaged can be culled, which will improve forest quality and structure. I take this approach for firewood harvesting. I don't get the best wood to burn, but I enjoy the aesthetics of a fire more than I need the heat from it. Those who depend on fires for heat will need more and higher-quality wood but can still harvest trees responsibly.

Wood as a construction material is a major usage of timber that will never go away. Market forces, however, are now changing how we use trees for the better. Our dependence on rapidly grown conifer wood for most construction has made larger timbers more difficult and costlier to produce. Frequent harvests yield smaller trees, producing smaller timber. Engineered wood products, formerly just used in specialized applications, are becoming commonplace. In these products, smaller pieces of wood, often from smaller trees, are glued together, resulting in solid framing members. Engineers can specify these components for most any length and application. They are straighter than traditional lumber, resist warping, and are quality controlled in a way impossible with unadulterated, natural wood. The earliest of these engineered products was plywood; it is now so accepted that we often do not consider it an engineered product. Newer products include sheathing (wall) and decking (floor) panels, wooden I-beams to replace floor joists, and siding options. Engineered products use more of the tree more efficiently and can use smaller trees useless for larger framing needs. The better quality of the engineered products and their decreasing cost relative to solid wood will make engineered products an increasing portion of the wood market. Engineered lumber requires fewer trees and uses them more completely. Traditional sawmills that produce lumber are starting to take sawdust, considered a waste product, and compress it into wood pellets for heating stoves, improving utilization of harvested wood and potentially reducing the need for firewood. Some burn the sawdust directly to fuel the kilns used to dry the lumber.

Paper products are another major global usage of trees. However, as more work becomes paperless and records switch to digital formats, the need for paper will decrease dramatically. I cannot see this trend reversing itself. One paper product increasing in usage, particularly with the upswell in online shopping, is corrugated cardboard and other packaging materials. Cardboard can be engineered into a fantastic array of shapes to ship almost everything safely without plastics. It is light, incredibly recyclable, and, if necessary, compostable. The ability to recycle cardboard means harvesting fewer trees. Even reusing cardboard fibers once halves the need for introducing new timber into the supply chain. The monetary value of cardboard to recyclers presents a financial motivation that has generated a network of businesses that improve the efficiency and rate of recycling.

The one place where market forces may not be sufficient to reduce wood use is as a disposable shipping product. Wooden pallets are commonplace and, in many situations, are used only once. Where shipping regularly goes between a central warehouse and individual locations, pallets are reused until they are no longer functional—this is an excellent system that does not need alteration. The problem arises from shipping that ends at the consumer with no mechanism to return the pallets to the shipping network. When we remodeled our current house, each appliance arrived on a pallet. I repurposed four of these as a compost bin, since rotted, but await a purpose for the others. Stores regularly have piles of pallets outside free for the taking. There are now books on crafts using pallet wood, so they are an increasingly available resource. Pallets would seem to be an excellent place for a deposit system. Deposits are common in battery recycling, drink bottles in many states, and kegs of beer. Such a system may encourage pallet makers to standardize sizes and make the pallets longer lasting, perhaps even from recycled plastic.

There are applications for which wood packaging is necessary for shipping. Large, manufactured parts that cannot be boxed often get strapped to platforms, crated, or enclosed in a wood frame that allows them to be tied down on a truck bed. There is no way to avoid this application, but there must be a way to reduce this wood from ending up in landfills. For years, my father would bring home 3 × 3 lumber that had been placed between layers of reinforcing steel that arrived on a construction site; their ultimate fate was the dumpster. We cut them into 18-inch (46-cm) pieces and burned them in our woodstove—it was the tightest stacked woodpile ever. Pass any construction site, and you will find a dumpster of discarded wood, used once and then abandoned. Manufacturers could recycle this material into engineered products, corrugated cardboard, or anything that would keep it out of a landfill. These and other efforts to recycle wood waste would significantly alleviate pressure on forests.

## Strategy 4: Manage and Monitor Our Forests

For many, forest management is synonymous with tree harvesting and focuses on activities intended to maximize tree growth and enhance the production of marketable timber. This management view is an agricultural approach designed to maximize profit and yield over time. Foresters in the past would often approach landowners with the ominous-sounding diagnosis that their forest was overmature. This declaration was, of course, accompanied by an offer to harvest the trees to get the system back to productivity. Overmaturity is simply when trees grow below their maximum rate, and wood accumulation decreases. As forests regrow after clearing, the trees grow rapidly at first, slowing as the trees get larger because of competition between trees and the increasing costs of an ever-enlarging tree body. More photosynthetic products need to be expended just to maintain a larger tree, a cost that is minimal in small trees. Reductions in wood production occur well before the forest would be considered mature in an ecological sense. Ironically, wood quality is poorer in more rapidly growing trees for most species.

While a yield-based approach to forest management may have been appropriate in the past, the ecological insights that have developed over the past few decades have generated additional management goals. We have broadened our forest perspective to encompass all species, not just trees, and to include metrics other than timber yield. Management planning often includes plant and animal species, exotic species, water quality, nutrient retention, carbon storage, and sustainability. These are management targets never imagined by early foresters. Both the U.S. and Canadian Forest Services manage vast tracts of land, facilitating tree harvests in addition to thinning forest stands, controlling exotic plant invaders, conducting controlled burns for the forests that require them, and planting trees when required. Both services have highly trained personnel to do this work and conduct research to continually refine management recommendations. Their work is critical, and, if anything, more foresters are needed to monitor and manage the vast areas involved effectively.

Federal agencies are responsible for much of North America's forests, but many forests are in private hands, particularly in the United States. Commercial companies manage large tracts, sometimes with a financial drive to focus purely on the bottom line. Happily, such a perspective encourages healthy trees and minimizes exotic species populations. Sadly, this perspective often ignores the forest's other residents and the benefits that they can bring. It is critically important that commercial forests are actively managed, have well-articulated goals, and have workers to achieve those goals. Hand in hand with such a management goal is monitoring to assess forest conditions and identify problems. Commercial management is not bad, although we may not all appreciate the forests that a financial perspective produces.

There are a multitude of small- to medium-sized forest parcels in local ownership that have commercial intentions or management plans. Some of these parcels may be harvested for timber to provide income for the owner, but often these forests are just left to their own devices. Most landowners do not have forestry management experience or training. Many are not able to differentiate native from exotic species and may even favor the non-native invaders. The land I grew up on became infested with bush honeysuckle (*Lonicera maackii*, chapter 7), which my father thought was pretty and fed the birds, much to my chagrin. People often own forested land for hunting or other recreational activities. Many absentee landowners may visit their forests only during hunting season. In my county in east-central Illinois, within the township that I dwell (36 square miles; 93 km²), we have forested land owned by people in Indiana (not surprising), Florida, Michigan, Ohio, Texas, Virginia, and Washington. These landowners cannot be carrying out any forest management—a benign neglect approach to land stewardship. Unfortunately, this is the approach that many landowners take.

Most forests in private hands have been disturbed by one or more bouts of tree harvesting in their history. As forests regrow, they start as very dense stands, thinning as individual trees die under the heavy burden of competition. These forests often have uncontrolled exotic plant invasions, retarding tree growth and slowing forest development for decades. Even native species such as grapes (*Vitis* spp.) can be problematic in disturbed forests. Any tree harvesting, even individual trees for firewood, can exacerbate these issues. Forest harvests on private lands are not closely regulated, and unless a landowner knows what to ask commercial logging groups, they receive a check and a property left in shambles. I have seen more than one reasonable forest converted to an impenetrable thicket of trees and shrubs by irresponsible logging practices. Unfortunately, many landowners do not know that they can expect better.

Young and high-density forests are perfect places for the issues described in the preceding chapters to occur. Pathogen transfer and pest movements are easier whenever the host tree occurs in high densities. As trees in dense forests often grow poorly and are continually stressed, they have increased chances of infection or susceptibility to an insect pest. Stressed plants have lower defenses, not unlike a human's immunological suppression by continued stress leads to increased illness. One of the common recommendations to mitigate forest pests is to maintain tree health, linked with broader spacing between trees. Invasions of exotic species often become established in forests disturbed by logging. If left unchecked, they may affect the forest in a multitude of ways. Problematically, the invaded forest then becomes a source of the invader to spread into other forests.

To me, the critical management issue for private landowners is monitoring by knowledgeable, unbiased people willing to help achieve a landowner's management goals. Governmental and commercial entities monitor their lands to prevent issues from expanding into problems. State or provincial and local agencies, with federal support, can provide experts willing to help individual landholders develop management plans for their forest. A part of this process is typically a site visit to identify issues and make recommendations. Of course, management planning also assumes that recommendations will be followed over the lifetime of a forest, effectively in perpetuity. Tree thinning, culling of undesirable species, and exotic species management are hard work, even more so if large acreages are involved. Once begun, continued maintenance can be much more manageable, as long as it does not cease altogether.

Unfortunately, programs to develop forest management plans have not been used by many landowners. Forest management is often perceived as unnecessary, an unwelcome intrusion, or the first step toward logging. Knowledge of the benefits arising from these programs lags dramatically behind other conservation initiatives, such as conservation reserve programs, further limiting adoption. Closing the monitoring gaps in forested landscapes is critical to our forests' future. Otherwise, the large forest areas not being adequately monitored represent a continual opportunity for the next forest pest or pathogen to become established undetected and perhaps lead to the loss of yet another tree species.

## The Elephants in the Room

Preventing and managing forest pests, maximizing forest health, and other pro-forest activities occur within the larger context of global climate change and nutrient deposition. These forces are dramatically altering the interaction landscape within which our forests and trees exist. Management practices employed today may not be suitable in a decade or two. Insect pests with limits imposed by winter temperatures may move northward with increasingly warming temperatures. Droughts may cause an environmental shrinkage of a tree species' habitat; this shift may be rapid, following an extreme drought, or slow if the environmental change is also slow. The accumulation of soil fertility by continued nitrogen deposition may cross a threshold, shifting the competitive balance from a native to a non-native tree. These phenomena may occur in the not-too-distant future; all pose potentially nightmarish management scenarios.

People, particularly governmental people, are planners. We like 5- or 10-year plans, or longer, for budgeting and assessing our successes. One of the most common job interview questions is, "Where do you see yourself in 5 years?" We ask this to see the trajectory a candidate believes they are on. The problem with planning in an ever-changing environmental landscape is that, though the target may remain the same, the way to get there is constantly shifting. Our planning, then, must also be revisable to account for unexpected changes and nimble enough to do so in a relevant time scale. This process is known as adaptive management, characterized by continual assessment and revision of goals and management strategies. This process must be supported by sufficient data that evaluate the system's current state to be effective. The adaptive nature of the approach is where issues arise. Sometimes planners do not like to update their approach, perhaps fearing that public opinion will consider their decisions to be inconsistent. Sometimes the motivation and funds for continued monitoring wane, leaving less feedback to fine-tune management efforts. We have seen this effect in the decades-long decline in public health funding, now regretted in the wake of COVID-19.

Climate change and nutrient deposition are not quickly addressed and would take decades of recovery even if stopped tomorrow. These issues touch on forests and are genuinely planetary in scale. In the short term, we should expect species distributions to change. We have historical examples of species' responses to global climate changes in the northward shifting of species as glaciers retreated. This change happened without human intervention, which is somewhat comforting in that it lends hope for a resilient transition in the climatic shift we are currently undergoing. However, two things dramatically differ now compared with the last glacial period. First, recent climate changes are happening much faster than in the post-glacial period. Second, we have converted large areas to agriculture or other nonnatural landscapes. These areas are barriers to plant species movements. Tree and forest species with good dispersal ability may move fast enough and cross such broad dispersal barriers. A great many species will not be able to achieve this. For those species, assisted migration—in this case, active movement of seeds or plants—will need to happen. That is a societal discussion that we are nowhere near ready to have. Selecting the species to move and which to let fend for themselves will almost certainly be contentious, as will a decision to do nothing.

Acid rain and nitrogen deposition associated with fossil fuel burning alter the foundations of all terrestrial systems, not just forests and the trees they contain. Alteration of soil pH, particularly over long periods as the capacity of soils to buffer the acid becomes depleted, will alter everything from nutrient retention to soil microbes and within-soil chemical processes. In addition, the increase in biologically available nitrogen will shift plant species from those able to compete for nitrogen to those best suited to bring in nitrogen when it is freely available. There are many native plants with this strategy, but there are also many non-native plants that appear to be particularly well suited to those conditions. As nitrogen in the soil passes to the plants and then is transmitted up the food web, there will likely be effects on other trophic levels that will be much more difficult to predict. Herbivores often respond to the level of nitrogen in plant tissues, so that we may expect an increase in native and non-native herbivores. It is almost impossible to project how insect feeding activity will affect the plant community, their predators, and other ecosystem components.

Such global issues form the context for the pathogens and pest insects currently in North America's forests. They also represent new stressors on our forests that may limit our ability to manage the exotic invasive plants, insects, and pathogens already established. There may be forest pests that have been unsuccessful in the past but can now establish and spread. Unfortunately, these global issues are much more complicated than individual forest pests and require a global solution.

## Victory within Our Grasp

I have been asked several times whether I am optimistic about the future of our forests. It feels somewhat odd, but honestly, I am optimistic. All too often, we see paths forward as all or nothing approaches. That view leaves only success and failure as options. I always think in gradients. Between success and failure lie better and worse, much more achievable in the short term. We may not intercept all forest pests at our borders, but if we can reduce them, we can do a lot to minimize impacts on forests. This is a better outcome. With a smaller burden of new challenges, we will also be able to address the current issues in our forests—also better. There are a lot of very clever researchers who have devoted their lives to understanding forest pests, developing biological controls, and engaging in other work necessary to achieve any level of success. I have had the good fortune to work with some of them, and I have faith in their abilities.

Science has generated a fabulous diversity of new ways to manipulate DNA and build resistance into trees within the last decade. We have seen massive increases in our ability to detect new invaders and, if motivated, we can move incredibly quickly to deal with them. If anything, the knowledge on each tree's challenges and the impacts of their losses are testimonials to the ability of science to address a broad range of questions. That is inspiring to me. Vigilance, monitoring, and research are not cheap, and they take people and time, but the rewards are tangible to anyone. Education and outreach are necessary for these efforts to ensure continued public support. All of this is for the better.

As the population of the planet becomes more urbanized, experiences with forests and trees will also decline. The environmental, psychological, and financial benefits of trees in urban areas are clear. We need to make sure that all neighborhoods have trees, not just the wealthy areas. Aldo Leopold based his land ethic on the central idea that we would preserve what we appreciate. I grew up with trees, at least after the age of six. I cannot imagine my childhood without trees as they were so central to my daily life. Most urban areas cannot sustain full forests, but a diversity of trees could and should be planted. Urban trees often do not live as long as their rural counterparts and often require more maintenance. However, more trees are always better, even if they are smaller or less healthy than they could be. The more people appreciate trees, the more people will appreciate forests, the greater the support for healthy native forests. Greater appreciation of forests is better in the long run.

My last cause for optimism comes from the trees themselves. A healthy and vigorous tree is easy to appreciate, stately and tall in the landscape. However, a tree that has struggled in its life forever bears the marks of that struggle on its limbs. A branch, once dead, never regrows, making a lasting impact on the tree's body. Bonsai, the Japanese art of miniaturizing trees, often focuses on generating the impression of age by capturing the struggle for existence in the plant. We are naturally drawn to this and interpret it as beauty. We appreciate the struggle captured artistically. We can see similar things in nature. Trees struck by lightning bear those scars, as do trees at high elevations where blowing ice has pruned away branches and bark. So many times, I have seen a struggling tree and thought that it could not possibly survive, but then I find a few green leaves and watch the tree grow back, a little bigger each passing year. The tenacity with which trees cling to life is inspirational. We still have surviving American elms and chestnuts decades after their struggle with disease began. They await a little help from us, those who caused their problems. I expect similar things with eastern hemlock and ash trees. We know the problems and are working diligently on solutions.

That, to me, is better.

# References by Chapter

## General Tree and Forest References

Braun, E.L. 1950. Deciduous forests of eastern North America. Philadelphia: Blakiston.

Peattie, D.C. 1948. A natural history of North American trees. San Antonio, TX: Trinity University Press.

Pinchot, G. 1903. A primer of forestry. Part 1. The forest. U.S. Department of Agriculture, Division of Forestry, Bulletin 24.

Pinchot, G. 1905. A primer of forestry. Part 2. Practical forestry. U.S. Department of Agriculture, Division of Forestry, Bulletin 24.

Thomas, P.A. 2014. Trees: Their natural history. 2nd ed. Cambridge: Cambridge University Press.

Tudge, C. 2005. The tree: A natural history of what trees are, how they live, and why they matter. New York: Three Rivers Press.

## Introduction: First, Some Context

Aukema, J.E., D.G. McCullough, B. von Holle, A.M. Liebhold, K. Britton, and S.J. Frankel. 2010. Historical accumulation of nonindigenous forest pests in the continental United States. BioScience 60:886–897.

Balding, M., and K.J.H. Williams. 2016. Plant blindness and the implications for plant conservation. Conservation Biology 30:1192–1199.

Burns, R.M., and B.H. Honkala, technical coordinators. 1990. Silvics of North America. Vol. 1. Conifers. Washington, DC: U.S. Department of Agriculture, Forest Service, Agriculture Handbook 654.

Burns, R.M., and B.H. Honkala, technical coordinators. 1990. Silvics of North America. Vol. 2. Hardwoods. Washington, DC: U.S. Department of Agriculture, Forest Service, Agriculture Handbook 654.

Jezkova, T., and J.J. Wiens. 2016. Rates of change in climatic niches in plant and animal populations are much slower than projected climate change. Proceedings of the Royal Society of London B: Biological Sciences 283:20162104.

Liebhold, A.M., D.G. Mccullough, L.M. Blackburn, S.J. Frankel, B. von Holle, and J.E. Aukema. 2013. A highly aggregated geographical distribution of forest pest invasions in the USA. Diversity and Distributions 19:1208–1216.

Lovett, G.M., M. Weiss, A.M. Liebhold, T.P. Holmes, B. Leung, K.F. Lambert, D.A. Orwig, et al. 2016. Nonnative forest insects and pathogens in the United States: Impacts and policy options. Ecological Applications 26:1437–1455.

## American Elm—*Ulmus americana*

Barnes, B.V. 1976. Succession in deciduous swamp communities of southeastern Michigan formerly dominated by American elm. Canadian Journal of Botany. 54:19–24.

Brasier, C.M. 1990. China and the origins of Dutch elm disease: An appraisal. Plant Pathology 39:5–16.

Brasier, C.M., and K.W. Buck. 2001. Rapid evolutionary changes in a globally invading fungal pathogen (Dutch elm disease). Biological Invasions 3:223–233.

Dunn, C.P. 1986. Shrub layer response to death of *Ulmus americana* in southeastern Wisconsin lowland forests. Bulletin of the Torrey Botanical Club 113:142–148.

Durrell, L. 1981. Memories of E. Lucy Braun. Ohio Biological Survey Biological Notes 15:37–39.

Huenneke, L.F. 1983. Understory response to gaps caused by the death of *Ulmus americana* in central New York. Bulletin of the Torrey Botanical Club 110:170–175.

Knight, K.S., J.M Slavicek, R. Kappler, B. Wiggin, and K. Menard. 2011. Using Dutch elm disease-tolerant elm to restore floodplains impacted by emerald ash borer. Proceedings of the 4th International Workshop on Genetics of Host–Parasite Interactions in Forestry, 317–323. U.S. Department of Agriculture, Forest Service, General Technical Report PSW-GTR-240.

Lanier, G.N., D.C. Schubert, and P.D. Manion. 1988. Dutch elm disease and elm yellows in central New York: Out of the frying pan into the fire. Plant Disease 72:189–194.

Marcone, C. 2016. Elm yellows: A phytoplasma disease of concern in forest and landscape ecosystems. Forest Pathology 47:1–13.

Martín, J.A., J. Witzell, K. Blumenstein, E. Rozpedowska, M. Helander, T.N. Sieber, and L. Gil. 2013. Resistance to Dutch elm disease reduces presence of xylem endophytic fungi in elms (*Ulmus* spp.). PLoS ONE 8(2): e56987.

Parker, G.R., and D.J. Leopold. 1983. Replacement of *Ulmus americana* L. in a mature east-central Indiana woods. Bulletin of the Torrey Botanical Club 110:482–488.

Richardson, C.J., and C.W. Cares. 2011. An analysis of elm (*Ulmus americana*) mortality in a second-growth hardwood forest in southeastern Michigan. Canadian Journal of Botany. 54:1120–1125.

Santini, A., and M. Faccoli. 2015. Dutch elm disease and elm bark beetles: A century of association. IForest–Biogeosciences and Forestry 8(2): 126–134.

Sinclair, W.A., A.M. Townsend, H.M. Griffiths, and T.H. Whitlow. 2000. Responses of six Eurasian *Ulmus* cultivars to a North American elm yellows phytoplasma. Plant Disease 84:1266–1270.

Slavicek, J.M, and K.S. Knight. 2012. Generation of American elm trees with tolerance to Dutch elm disease through controlled crosses and selection. Proceedings of the 4th International Workshop on Genetics of Host–Parasite Interactions in Forestry, 342–346. U.S. Department of Agriculture, Forest Service, General Technical Report PSW-GTR-240.

Swingle, R.U., R.R. Whitten, and E.G. Brewer. 1949. Dutch elm disease. In: Trees: The yearbook of agriculture, 1949, ed. A. Stefferud. Washington, DC: U.S. Department of Agriculture.

## American Chestnut—*Castanea dentata*

Anagnostakis, S.L. 1982. Biological control of chestnut blight. Science 215:466–471.

Anagnostakis, S.L. 1987. Chestnut blight : The classical problem of an introduced pathogen. Mycologia 79:23–37.

Brewer, L.G. 1995. Ecology of survival and recovery from blight in American chestnut trees (*Castanea dentata* (Marsh.) Borkh.) in Michigan. Bulletin of the Torrey Botanical Club 122:40–57.

Dalgleish, H.J., C.D. Nelson, J.A. Scrivani, and D.F. Jacobs. 2016. Consequences of shifts in abundance and distribution of American chestnut for restoration of a foundation forest tree. Forests 7:1–9.

Dalgleish, H.J., and R.K. Swihart. 2012. American chestnut past and future: Implications of restoration for resource pulses and consumer populations of eastern U.S. forests. Restoration Ecology 20:490–497.

Elliott, K.J., and W.T. Swank. 2008. Long-term changes in forest composition and diversity following early logging (1919–1923) and the decline of American chestnut (*Castanea dentata*). Plant Ecology 197:155–172.

Faison, E.K., and D.R. Foster. 2014. Did American chestnut really dominate the eastern forest? Arnoldia 72:18–32.

Gilland, K.E., C.H. Keiffer, and B.C. McCarthy. 2012. Seed production of mature forest-grown American chestnut (*Castanea dentata* (Marsh.) Borkh). Journal of the Torrey Botanical Society 139:283–289.

Good, N.F. 2017. A study of natural replacement of chestnut in six stands in the highlands of New Jersey. Bulletin of the Torrey Botanical Club 95:240–253.

Heiniger, U., and D. Rigling. 1994. Biological control of chestnut blight in Europe. Annual Review of Phytopathology 32:581–599.

Jacobs, D.F., H.J. Dalgleish, and C.D. Nelson. 2013. A conceptual framework for restoration of threatened plants: The effective model of American chestnut (*Castanea dentata*) reintroduction. New Phytologist 197:378–393.

Keever, C. 1953. Present composition of some stands of the former oak–chestnut forest in the southern Blue Ridge Mountains. Ecology 34:44–54.

Latham, R.E. 1992. Co-occurring tree species change rank in seedling performance with resources varied experimentally. Ecology 73:2129–2144.

McCormick, J.F., and R.B. Platt. 1980. Recovery of an Appalachian forest following the chestnut blight or Catherine Keever—you were right! American Midland Naturalist 104:264–273.

McEwan, R.W., C.H. Keiffer, and B.C. McCarthy. 2006. Dendroecology of American chestnut in a disjunct stand of oak–chestnut forest. Canadian Journal of Forest Research. 36:1–11.

Milgroom, M.G., and P. Cortesi. 2004. Biological control of chestnut blight with hypovirulence: A critical analysis. Annual Review of Phytopathology 42:311–338.

Milgroom, M.G., K. Wang, Y. Zhou, and S.E. Lipari. 1996. Intercontinental population structure of the chestnut blight fungus, *Cryphonectria parasitica*. Mycologia. 88:179–190.

Paillet, F.L. 1982. The ecological significance of American chestnut (*Castanea dentata* (Marsh.) Borkh.) in the Holocene forests of Connecticut. Bulletin of the Torrey Botanical Club 109:457–473.

Paillet, F.L. 2002. Chestnut: History and ecology of a transformed species. Journal of Biogeography 29:1517–1530.

Paillet, F.L., and P.A. Rutter. 1989. Replacement of native oak and hickory tree species by the introduced American chestnut (*Castanea dentata*) in southwestern Wisconsin. Canadian Journal of Botany 67:3457–3469.

Russell, E.W.B. 1987. Pre-blight distribution of *Castanea dentata* (Marsh.) Borkh. Bulletin of the Torrey Botanical Club 114:183–190.

Springer, J.C., A.L. Davelos Baines, D.W. Fulbright, M.T. Chansler, and A.M. Jarosz. 2013. Hyperparasites influence population structure of the chestnut blight pathogen, *Cryphonectria parasitica*. Phytopathology 103:1280–1286.

Steiner, K.C., J.W. Westbrook, F.V. Hebard, L.L. Georgi, W.A. Powell, and S.F. Fitzsimmons. 2016. Rescue of American chestnut with extraspecific genes following its destruction by a naturalized pathogen. New Forests 48:317–336.

Whittaker, R.H. 1956. Vegetation of the Great Smoky Mountains. Ecological Monographs 26:2–80.

### Eastern Hemlock—*Tsuga canadensis*

Albani, M., P.R. Moorcroft, A.M. Ellison, D.A. Orwig, and D.R. Foster. 2010. Predicting the impact of hemlock woolly adelgid on carbon dynamics of Eastern United States forests. Canadian Journal of Forest Research 40:119–133.

Allison, T.D., R.E. Moeller, and M.B. Davis. 1986. Pollen in laminated sediments provides evidence for a mid-Holocene forest pathogen outbreak. Ecology 67:1101–1105.

Cheah, C., M.E. Montgomery, S. Salom, B.L. Parker, S. Costa, and M. Skinner. 2004. Biological control of hemlock woolly adelgid. U.S. Department of Agriculture, Forest Service, Forest Health Technology Enterprise Team.

Eschtruth, A.K., N.L. Cleavitt, J.J. Battles, R.A. Evans, and T.J. Fahey. 2006. Vegetation dynamics in declining eastern hemlock stands: 9 years of forest response to hemlock woolly adelgid infestation. Canadian Journal of Forest Research 36:1435–1450.

Foster, D.R., W.W. Oswald, E.K. Faison, E.D. Doughty, and B.C.S. Hansen. 2007. A climatic driver for abrupt mid-Holocene vegetation dynamics and the hemlock decline in New England. Ecology. 87:2959–2966.

Havill, N.P., M.E. Montgomery, G. Yu, S. Shiyake, and A. Caccone. 2006. Mitochondrial DNA from hemlock woolly adelgid (Hemiptera: Adelgidae) suggests cryptic speciation and pinpoints the source of the introduction to eastern North America. Annals of the Entomological Society of America 99:195–203.

Jenkins, J.C., J.D. Aber, and C.D. Canham. 1999. Hemlock woolly adelgid impacts on community structure and N cycling rates in eastern hemlock forests. Canadian Journal of Forest Research 29:630–645.

Louda, S.M., D. Kendall, J. Connor, and D. Simberloff. 1997. Ecological effects of an insect introduced for the biological control of weeds. Science 277:1088–1090.

McClure, M.S. 1987. Biology and control of hemlock woolly adelgid. Bulletin—Connecticut Agricultural Experiment Station 851.

McClure, M.S., and C.A.S.J. Cheah. 1999. Reshaping the ecology of invading populations of hemlock woolly adelgid, Adelges tsugae (Homoptera: Adelgidae), in eastern North America. Biological Invasions 1:247–254.

Orwig, D.A., R.C. Cobb, A.W. D'Amato, M.L. Kizlinski, and D.R. Foster. 2008. Multi-year ecosystem response to hemlock woolly adelgid infestation in southern New England forests. Canadian Journal of Forest Research 38:834–843.

Orwig, D.A., and D.R. Foster. 2016. Forest response to the introduced hemlock woolly adelgid in southern New England, USA. Journal of the Torrey Botanical Society 125:60–73.

Orwig, D.A., D.R. Foster, and D.L. Mauselt. 2002. Landscape patterns of hemlock decline in New England due to the introduced hemlock woolly adelgid. Journal of Biogeography 29:1475–1487.

Oswald, W.W., E.D. Doughty, D.R. Foster, B.N. Schuman, and D.L.Wagner. 2017. Evaluating the role of insects in the middle-Holocene Tsuga decline. Journal of the Torrey Botanical Society 144:35–39.

Ross, R.M., R.M. Bennett, C.D. Snyder, J.A. Young, D.R. Smith, and D.P. Lemarié. 2003. Influence of eastern hemlock (Tsuga canadensis L.) on fish community structure and function in headwater streams of the Delaware River basin. Ecology of Freshwater Fish 12:60–65.

Royle, D.D., and R.G. Lathrop. 2002. Discriminating Tsuga canadensis hemlock forest defoliation using remotely sensed change detection. Journal of Nematology 34:213–221.

Skinner, M., B.L. Parker, S. Gouli, and T. Ashikaga. 2003. Regional responses of hemlock woolly adelgid (Homoptera: Adelgidae) to low temperatures. Environmental Entomology 32:523–528.

Spaulding, H.L., and L.K. Rieske. 2010. The aftermath of an invasion: Structure and composition of central Appalachian hemlock forests following establishment of the hemlock woolly adelgid, Adelges tsugae. Biological Invasions 12:3135–3143.

Stadler, B., T. Müller, and D. Orwig. 2009. The ecology of energy and nutrient fluxes in hemlock forests invaded by hemlock woolly adelgid. Ecology 87:1792–1804.

Stadler, B., T. Müller, D. Orwig, and R. Cobb. 2005. Hemlock woolly adelgid in New England forest: Canopy impacts transforming ecosystem processes and landscapes. Ecosystems 8:233–247.

Tighe, M.E., W.S. Dvorak, W.A. Whittier, J.L. Romero, and J.R. Rhea. 2005. The ex situ conservation of Carolina hemlock. In: Third Symposium on Hemlock Woolly Adelgid in the Eastern United States, ed. B. Onken and R. Reardon, 180–190. U.S. Department of Agriculture, Forest Service, Forest Health Technology Enterprise Team.

Tingley, M.W., D.A. Orwig, R. Field, G. Motzkin, M.W. Tingley, D.A. Orwig, R. Field, and G. Motzkin. 2017. Avian response to removal of a forest dominant: consequences of hemlock woolly adelgid infestations. Journal of Biogeography 29:1505–1516.

Wallace, M.S., and F.P. Hain. 2000. Field surveys and evaluation of native and established predators of the hemlock woolly adelgid (Homoptera: Adelgidae) in the southeastern United States. Environmental Entomology 29:638–644.

Webster, J.R., K. Morkeski, C.A. Wojculewski, B.R. Niederlehner, F. Benfield, and K.J. Elliott. 2012. Effects of hemlock mortality on streams in the southern Appalachian Mountains. American Midland Naturalist 168:112–131.

## White Ash—*Fraxinus americana*

Abell, K., T. Poland, A. Cossé, and L. Bauer. 2015. Trapping techniques for emerald ash borer and its introduced parasitoids. In: Biology and control of emerald ash borer, ed. R.G. Van Driesche and R.C. Reardon, 113–127. U.S. Department of Agriculture, Forest Service, Forest Health Technology Enterprise Team.

Abell, K.J., L.S. Bauer, J.J. Duan, and R. Van Driesche. 2014. Long-term monitoring of the introduced emerald ash borer (Coleoptera: Buprestidae) egg parasitoid, *Oobius agrili* (Hymenoptera: Encyrtidae), in Michigan, USA and evaluation of a newly developed monitoring technique. Biological Control 79:36–42.

Abella, S.R., C.E. Hausman, J.F. Jaeger, K.S. Menard, T.A. Schetter, and O.J. Rocha. 2019. Fourteen years of swamp forest change from the onset, during, and after invasion of emerald ash borer. Biological Invasions 21:3685–3696.

Bauer, L.S., J.J. Duan, J.R. Gould, and R. Van Driesche. 2015. Progress in the classical biological control of *Agrilus planipennis* Fairmaire (Coleoptera: Buprestidae) in North America. Canadian Entomologist 147:300–317.

Bauer, L.S., J.J. Duan, J.P. Lelito, H. Liu, and J.R. Gould. 2015. Biology of emerald ash borer parasitoids. In: Biology and control of emerald ash borer, ed. R.G. Van Driesche and R.C. Reardon, 97–112. U.S. Department of Agriculture, Forest Service, Forest Health Technology Enterprise Team.

Boltz, B., and J. Wiedenbeck. 2010. Strike one! Aluminum. Strike two! Maple. Will EAB be strike three? In: 2010 Proceedings Symposium on Ash in North America, 26–31. U.S. Department of Agriculture, Forest Service, Northern Research Station GTR-NRS-P-72.

Burr, S.J., and D.G. McCullough. 2014. Condition of green ash (*Fraxinus pennsylvanica*) overstory and regeneration at three stages of the emerald ash borer invasion wave. Canadian Journal of Forest Research 44:768–776.

Cipollini, D. 2015. White fringetree as a novel larval host for emerald ash borer. Journal of Economic Entomology 108:370–375.

Cipollini, D., and D.L. Peterson. 2018. The potential for host switching via ecological fitting in the emerald ash borer–host plant system. Oecologia 187:507–519.

Crook, D.J., A. Khrimian, J.A. Francese, I. Fraser, T.M. Poland, A.J. Sawyer, and V.C. Mastro. 2008. Development of a host-based semiochemical lure for trapping emerald ash borer *Agrilus planipennis* (Coleoptera: Buprestidae). Environmental Entomology 37:356–365.

Duan, J.J., K.J. Abell, L.S. Bauer, J. Gould, and R. van Driesche. 2014. Natural enemies implicated in the regulation of an invasive pest: A life table analysis of the population dynamics of the emerald ash borer. Agricultural and Forest Entomology 16:406–416.

Duan, J.J., L.S. Bauer, K.J. Abell, M.D. Ulyshen, and R.G. van Driesche. 2015. Population dynamics of an invasive forest insect and associated natural enemies in the aftermath of invasion: Implications for biological control. Journal of Applied Ecology 52:1246–1254.

Duan, J.J., L.S. Bauer, K.J. Abell, and R. van Driesche. 2012. Population responses of hymenopteran parasitoids to the emerald ash borer (Coleoptera: Buprestidae) in recently invaded areas in north central United States. BioControl 57:199–209.

Duan, J.J., L.S. Bauer, and R.G. van Driesche. 2017. Emerald ash borer biocontrol in ash saplings: The potential for early stage recovery of North American ash trees. Forest Ecology and Management 394:64–72.

Flower, C.E., K.S. Knight, and M.A. Gonzalez-Meler. 2013. Impacts of the emerald ash borer (*Agrilus planipennis* Fairmaire) induced ash (*Fraxinus* spp.) mortality on forest carbon cycling and successional dynamics in the eastern United States. Biological Invasions 15:931–944.

Flower, C.E., K.S. Knight, J. Rebbeck, and M.A. Gonzalez-Meler. 2013. The relationship between the emerald ash borer (*Agrilus planipennis*) and ash (*Fraxinus* spp.) tree decline: Using visual canopy condition assessments and leaf isotope measurements to assess pest damage. Forest Ecology and Management 303:143–147.

Flower, C.E., L.C. Long, K.S. Knight, J. Rebbeck, J.S. Brown, M.A. Gonzalez-Meler, and C.J. Whelan. 2014. Native bark-foraging birds preferentially forage in infected ash (*Fraxinus* spp.) and prove effective predators of the invasive emerald ash borer (*Agrilus planipennis* Fairmaire). Forest Ecology and Management 313:300–306.

Haack, R.A., E. Jendek, H. Liu, K.R. Marchant, T.R. Petrice, T.M. Poland, and H. Ye. 2002. The emerald ash borer: A new exotic pest in North America. Michigan Entomological Society 47:1–5.

Hanberry, B.B. 2014. Rise of *Fraxinus* in the United States between 1968 and 2013. Journal of the Torrey Botanical Society 141:242–249.

Hauer, R.J., and W.D. Peterson. 2017. Effects of emerald ash borer on municipal forestry budgets. Landscape and Urban Planning 157:98–105.

Hoven, B.M., D.L. Gorchov, K.S. Knight, and V.E. Peters. 2017. The effect of emerald ash borer-caused tree mortality on the invasive shrub Amur honeysuckle and their combined effects on tree and shrub seedlings. Biological Invasions 19:2813–2836.

Klooster, W.S., D.A. Herms, K.S. Knight, C.P. Herms, D.G. McCullough, A. Smith, K.J.K. Gandhi, and J. Cardina. 2014. Ash (*Fraxinus* spp.) mortality, regeneration, and seed bank dynamics in mixed hardwood forests following invasion by emerald ash borer (*Agrilus planipennis*). Biological Invasions 16:859–873.

Knight, K.S., J.P. Brown, and R.P. Long. 2013. Factors affecting the survival of ash (*Fraxinus* spp.) trees infested by emerald ash borer (*Agrilus planipennis*). Biological Invasions 15:371–383.

Knight, K.S., D. Herms, R. Plumb, E. Sawyer, D. Spalink, E. Pisarczyk, B. Wiggin, R. Kappler, E. Ziegler, and K. Menard. 2012. Dynamics of surviving ash (*Fraxinus* spp.) populations in areas long infested by emerald ash borer (*Agrilus planipennis*). Proceedings of the 4th International Workshop on Genetics of Host–Parasite Interactions in Forestry, 143–152. U.S. Department of Agriculture, Forest Service, General Technical Report PSW-GTR-240.

Koenig, W.D., and A.M. Liebhold. 2017. A decade of emerald ash borer effects on regional woodpecker and nuthatch populations. Biological Invasions 19:2029–2037.

Kovacs, K.F., R.G. Haight, D.G. McCullough, R.J. Mercader, N.W. Siegert, and A.M. Liebhold. 2010. Cost of potential emerald ash borer damage in U.S. communities, 2009–2019. Ecological Economics 69:569–578.

MacFarlane, D.W., and S.P. Meyer. 2005. Characteristics and distribution of potential ash tree hosts for emerald ash borer. Forest Ecology and Management 213:15–24.

McKenney, D.W., J.H. Pedlar, D. Yemshanov, D.B. Lyons, K.L. Campbell, and K. Lawrence. 2012. Estimates of the potential cost of emerald ash borer (*Agrilus planipennis* Fairmaire) in Canadian municipalities. Arboriculture and Urban Forestry 38:81–91.

Muirhead, J.R., B. Leung, C. Van Overdijk, D.W. Kelly, K. Nandakumar, K.R. Marchant, and H.J. MacIsaac. 2006. Modelling local and long-distance dispersal of invasive emerald ash borer *Agrilus planipennis* (Coleoptera) in North America. Diversity and Distributions 12:71–79.

Orlova-Bienkowskaja, M.J., and M.G. Volkovitsh. 2018. Are native ranges of the most destructive invasive pests well known? A case study of the native range of the emerald ash borer, *Agrilus planipennis* (Coleoptera: Buprestidae). Biological Invasions 20:1275–1286.

Palik, B.J., A.W. D'Amato, and R.A. Slesak. 2021. Wide-spread vulnerability of black ash (*Fraxinus nigra* Marsh.) wetlands in Minnesota USA to loss of tree dominance from invasive emerald ash borer. Forestry 94:455–463.

Smith, A., D.A. Herms, R.P. Long, and K.J.K. Gandhi. 2015. Community composition and structure had no effect on forest susceptibility to invasion by the emerald ash borer (Coleoptera: Buprestidae). Canadian Entomologist 147:318–328.

Smitley, D., T. Davis, and E. Rebek. 2008. Progression of ash canopy thinning and dieback outward from the initial infestation of emerald ash borer (Coleoptera: Buprestidae) in southeastern Michigan. Journal of Economic Entomology 101:1643–1650.

## Sugar Maple—*Acer saccharum*

Abrams, M.D. 1992. Fire and the development of oak forests. BioScience 42:346–353.

Abrams, M.D. 1998. The red maple paradox: What explains the widespread expansion of red maple in eastern forests? BioScience 48:355–364.

Abrams, M.D. 2008. The demise of fire and 'mesophication' of forests in the Eastern United States. BioScience 58:123–138.

Bal, T.L., A.J. Storer, M.F. Jurgensen, P.V. Doskey, and M.C. Amacher. 2015. Nutrient stress predisposes and contributes to sugar maple dieback across its northern range: A review. Forestry 88:64–83.

Boggess, W.R. 1964. Trelease woods, Champaign County, Illinois: Woody vegetation and stand composition. Transactions of the Illinois Academy of Sciences 57:261–271.

Boggess, W.R., and L.W. Bailey. 1964. Brownfield Woods, Illinois: Woody vegetation and changes since 1925. American Midland Naturalist 71:392–401.

Canham, C.D. 1985. Suppression and release during canopy recruitment in *Acer saccharum*. Bulletin of the Torrey Botanical Club 112:134–145.

Carter, M., M. Smith, and R. Harrison. 2010. Genetic analyses of the Asian longhorned beetle (Coleoptera, Cerambycidae, *Anoplophora glabripennis*), in North America, Europe and Asia. Biological Invasions 12:1165–1182.

Cirelli, D., R. Jagels, and M.T. Tyree. 2008. Toward an improved model of maple sap exudation: The location and role of osmotic barriers in sugar maple, butternut and white birch. Tree Physiology 28:1145–1155.

Copenheaver, C.A., R.C. McCune, E.A. Sorensen, M.F.J. Pisaric, and B.J. Beale. 2014. Decreased radial growth in sugar maple trees tapped for maple syrup. Forestry Chronicle 90:771–777.

Dodds, K.J., and D.A. Orwig. 2011. An invasive urban forest pest invades natural environments—Asian long horned beetle in northeastern US hardwood forests. Canadian Journal of Forest Research 41:1729–1742.

Duan, J.J., E. Aparicio, D. Tatman, M.T. Smith, and D.G. Luster. 2016. Potential new associations of North American parasitoids with the invasive Asian longhorned beetle (Coleoptera: Cerambycidae) for biological control. Journal of Economic Entomology 109:699–704.

Hallett, R.A., S.W. Bailey, S.B. Horsley, and R.P. Long. 2006. Influence of nutrition and stress on sugar maple at a regional scale. Canadian Journal of Forest Research 36:2235–2246.

Horsley, S.B., R.P. Long, S.W. Bailey, R.A. Hallett, and T.J. Hall. 2000. Factors associated with the decline disease of sugar maple on the Allegheny Plateau. Canadian Journal of Forest Research 30:1365–1378.

Horsley, S.B., R.P. Long, S.W. Bailey, R.A. Hallett, and P.M. Wargo. 2002. Health of eastern North American sugar maple forests and factors affecting decline. Northern Journal of Applied Forestry 19:34–44.

Houston, D.R., and S.B. Horsley. 1999. History of sugar maple decline. In: Sugar maple ecology and health: Proceedings of an International Symposium, 19–26. Radnor, PA: U.S. Department of Agriculture, Forest Service, General Technical Report NE-261.

Hu, J., S. Angeli, S. Schuetz, Y. Luo, and A.E. Hajek. 2009. Ecology and management of exotic and endemic Asian longhorned beetle *Anoplophora glabripennis*. Agricultural and Forest Entomology 11:359–375.

MacLeod, A., H.F. Evans, and R.H.A. Baker. 2002. An analysis of pest risk from an Asian longhorn beetle (*Anoplophora glabripennis*) to hardwood trees in the European community. Crop Protection 21:635–645.

Marks, P.L., and S. Gardescu. 1998. A case study of sugar maple (*Acer saccharum*) as a forest seedling bank species. Journal of the Torrey Botanical Society 125:287–296.

Muhr, J., C. Messier, S. Delagrange, S. Trumbore, X. Xu, and H. Hartmann. 2016. How fresh is maple syrup? Sugar maple trees mobilize carbon stored several years previously during early springtime sap-ascent. New Phytologist 209:1410–1416.

Nowak, D.J., J.E. Pasek, R.A. Sequeira, D.E. Crane, and V.C. Mastro. 2001. Potential effect of *Anoplophora glabripennis* (Coleoptera: Cerambycidae) on urban trees in the United States. Journal of Economic Entomology 94:116–122.

Peterson, A.T., R. Scachetti-Pereira, and W.W. Hargrove. 2004. Potential geographic distribution of *Anoplophora glabripennis* (Coleoptera: Cerambycidae) in North America. American Midland Naturalist 151:170–178.

Poland, T.M. 1998. Chicago joins New York in battle with the Asian longhorned beetle. Newsletter of the Michigan Entomological Society 43:15–17.

Raupp, M.J., A.B. Cumming, and E.C. Raupp. 2006. Street tree diversity in eastern North America and its potential for tree loss to exotic borers. Journal ofArboriculture 32:297–304.

Smith, M.T., J. Bancroft, G. Li, R. Gao, and S. Teale. 2001. Dispersal of *Anoplophora glabripennis* (Cerambycidae). Environmental Entomology 30:1036–1040.

Smith, M.T., J. Bancroft, and J. Tropp. 2002. Age-specific fecundity of *Anoplophora glabripennis* (Coleoptera: Cerambycidae) on three tree species infested in the United States. Environmental Entomology 31:76–83.

Smith, M.T., J.J. Turgeon, P. De Groot, and B. Gasman. 2009. Asian longhorned beetle *Anoplophora glabripennis* (Motschulsky): Lessons learned and opportunities to improve the process of eradication and management. American Entomologist 55:21–25.

Sperry, J.S., J.R. Donnelly, and M.T. Tyree. 2008. Seasonal occurrence of xylem embolism in sugar maple (*Acer saccharum*). American Journal of Botany 75:1212–1218.

Williams, D.W., H.P. Lee, and I.K. Kim. 2004. Distribution and abundance of *Anoplophora glabripennis* (Coleoptera: Cerambycidae) in natural *Acer* stands in South Korea. Environmental Entomology 33:540–545.

Wong, B.L., K.L. Baggett, and A.H. Rye. 2003. Seasonal patterns of reserve and soluble carbohydrates in mature sugar maple (*Acer saccharum*). Canadian Journal of Botany 81:780–788.

## Other Trees with Other Challenges

Asaro, C., and L.A. Chamberlin. 2015. Outbreak history (1953–2014) of spring defoliators impacting oak-dominated forests in Virginia, with emphasis on gypsy moth (*Lymantria dispar* L.) and fall cankerworm (*Alsophila pometaria* Harris). American Entomologist 61:174–185.

Bellinger, R.G., F.W. Ravlin, and M.L. McManus. 1989. Forest edge effects and their influence on gypsy moth (Lepidoptera: Lymantriidae) egg mass distributions. Environmental Entomology 18:840–843.

Britton, K.O., W.D. Pepper, D.L. Loftis, and D.O. Chellemi. 1994. Effect of timber harvesting practices on polulations of *Cornus florida* and severity of dogwood anthracnose in western North Carolina. Plant Disease 78:398–402.

Cale, J.A., S.K. Letkowski, S.A. Teale, and J.D. Castello. 2012. Beech bark disease: An evaluation of the predisposition hypothesis in an aftermath forest. Forest Pathology 42:52–56.

Chellemi, D.O., and K.O. Britton. 1992. Influence of canopy microclimate on incidence and severity of dogwood anthracnose. Canadian Journal of Botany 70:1093–1096.

DiGregorio, L.M, M.E. Krasny, and T.J. Fahey. 1999. Radial growth trends of sugar maple (*Acer saccharum*) in an Allegheny northern hardwood forest affected by beech bark disease. Journal of the Torrey Botanical Society 126:245–254.

Duchesne, L., and R. Ouimet. 2009. Present-day expansion of American beech in north-eastern hardwood forests: Does soil base status matter? Canadian Journal of Forest Research 39:2273–2282.

Eisenbies, M.H., C. Davidson, J. Johnson, R. Amateis, and K. Gottschalk. 2007. Tree mortality in mixed pine–hardwood stands defoliated by the European gypsy moth (*Lymantria dispar* L.). Forest Science 53:683–691.

Forrester, J.A., G.G. Mcgee, and M.J. Mitchell. 2003. Effects of beech bark disease on aboveground biomass and species composition in a mature northern hardwood forest, 1985 to 2000. Journal of the Torrey Botanical Society 130:70–78.

Foss, L.K., and L.K. Rieske. 2003. Species-specific differences in oak foliage affect preference and performance of gypsy moth caterpillars. Entomologia Experimentalis et Applicata 108:87–93.

Grant, J.F., M.T. Windham, W.G. Haun, G.J. Wiggins, and P.L. Lambdin. 2011. Initial assessment of thousand cankers disease on black walnut, *Juglans nigra*, in eastern Tennessee. Forests 2:741–748.

Griffin, G.J. 2015. Status of thousand cankers disease on eastern black walnut in the eastern United States at two locations over 3 years. Forest Pathology 45:203–214.

Griffin, J.M., G.M. Lovett, M.A. Arthur, and K.C. Weathers. 2003. The distribution and severity of beech bark disease in the Catskill Mountains, N.Y. Canadian Journal of Forest Research 33:1754–1760.

Hadziabdic, D., L.M. Vito, M.T. Windham, J.W. Pscheidt, R.N. Trigiano, and M. Kolarik. 2014. Genetic differentiation and spatial structure of *Geosmithia morbida*, the causal agent of thousand cankers disease in black walnut (*Juglans nigra*). Current Genetics 60:75–87.

Hajek, A.E., and P.C. Tobin. 2008. North American eradications of Asian and European gypsy moth. In: Use of microbes for control and eradication of invasive arthropods, ed. A.E. Hajek, T.R. Glare, and M. O'Callaghan, 71–89. Dordrecht: Springer.

Hancock, J.E., M.A. Arthur, K.C. Weathers, and G.M. Lovett. 2008. Carbon cycling along a gradient of beech bark disease impact in the Catskill Mountains, New York. Canadian Journal of Forest Research 38:1267–1274.

Hiers, J.K., and J.P. Evans. 1997. Effects of anthracnose on dogwood mortality and forest composition of the Cumberland Plateau (U.S.A.). Conservation Biology 11:1430–1435.

Holzmueller, E.J., S. Jose, and M.A. Jenkins. 2008. The relationship between fire history and an exotic fungal disease in a deciduous forest. Oecologia 155:347–356.

Holzmueller, E., S. Jose, M. Jenkins, A. Camp, and A. Long. 2006. Dogwood anthracnose in eastern hardwood forests: What is known and what can be done? Journal of Forestry 104:21–26.

Jedlicka, J., J. Vandermeer, K. Aviles-Vazquez, O. Barros, and I. Perfecto. 2004. Gypsy moth defoliation of oak trees and a positive response of red maple and black cherry: An example of indirect interaction. American Midland Naturalist 152:231–236.

Jenkins, M.A., and P.S. White. 2018. *Cornus florida* L. mortality and understory composition changes in western Great Smoky Mountains National Park. Journal of the Torrey Botanical Society 129:194–206.

Koch, J.L. 2010. Beech bark disease: The oldest 'new' threat to American beech in the United States. Outlooks on Pest Management 21:64–68.

Latty, E.F., C.D. Canham, and P.L. Marks. 2003. Beech bark disease in northern hardwood forests: The importance of nitrogen dynamics and forest history for disease severity. Canadian Journal of Forest Research 33:257–268.

Leak, W.B. 2006. Fifty-year impacts of the beech bark disease in the Bartlett Experimental Forest, New Hampshire. Northern Journal of Applied Forestry 23:141–143.

Liebhold, A., J. Elkinton, D. Williams, and R.M. Muzika. 2000. What causes outbreaks of the gypsy moth in North America? Population Ecology 42:257–266.

Liebhold, A.M., J.A. Halverson, and G.A. Elmes. 1992. Gypsy moth invasion in North America: A quantitative analysis. Journal of Biogeography 19:513–520.

LoGiudice, K., R.S. Ostfeld, K.A. Schmidt, and F. Keesing. 2003. The ecology of infectious disease: Effects of host diversity and community composition on Lyme disease risk. Proceedings of the National Academy of Sciences 100:567–571.

McEwan, R.W., R.N. Muller, M.A. Arthur, and H.H. Housman. 2000. Temporal and ecological patterns of flowering dogwood mortality in the mixed mesophytic forest of eastern Kentucky. Journal of the Torrey Botanical Society 127:221–229.

Morin, R.S., A.M. Liebhold, P.C. Tobin, K.W. Gottschalk, and E. Luzader. 2007. Spread of beech bark disease in the eastern United States and its relationship to regional forest composition. Canadian Journal of Forest Research 37:726–736.

Ostfeld, R.S. 2011. Lyme disease: The ecology of a complex system. New York: Oxford University Press.

Ostfeld, R.S., C.G. Jones, and J.O. Wolff. 1996. Of mice and mast. BioScience 46:323–330.

Papaik, M.J., C.D. Canham, E.F. Latty, and K.D. Woods. 2005. Effects of an introduced pathogen on resistance to natural disturbance: Beech bark disease and windthrow. Canadian Journal of Forest Research 35:1832–1843.

Pasquarella, V.J., J.S. Elkinton, and B.A. Bradley. 2018. Extensive gypsy moth defoliation in southern New England characterized using landsat satellite observations. Biological Invasions 20:3047–3053.

Pierce, A.R., W.R. Bromer, and K.N. Rabenold. 2008. Decline of *Cornus florida* and forest succession in a *Quercus–Carya* forest. Plant Ecology 195:45–53.

Randolph, K.C., A.K. Rose, C.M. Oswalt, and M.J. Brown. 2013. Status of black walnut (*Juglans nigra* L.) in the eastern United States in light of the discovery of thousand cankers disease. *Castanea* 78:2–14.

Redlin, S.C. 1991. *Discula destructiva* sp. nov., cause of dogwood anthracnose. Mycologia 83:633–642.

Rugman-Jones, P.F., S.J. Seybold, A.D. Graves, and R. Stouthamer. 2015. Phylogeography of the walnut twig beetle, *Pityophthorus juglandis*, the vector of thousand cankers disease in North American walnut trees. PLoS ONE 10(2): e0118264.

Seybold, S.J., P.L. Dallara, S.M. Hishinuma, and M.L. Flint,. 2013. Detecting and identifying walnut twig beetle: Monitoring guidelines for the invasive vector of thousand cankers disease of walnut. University of California Division of Agriculture and Natural Resources, UC IPM Program.

Sharov, A.A., D.S. Leonard, A.M. Liebhold, E.A. Roberts, and W. Dickerson. 2002. 'Slow the spread' a national program to contain the gypsy moth. Journal of Forestry 100:30–36.

Sherald, J.L, T.M. Stidham, J.M. Hadidian, and J.E. Hoeldke. 1996. Progression of the dogwood anthracnose epidemic and the status of flowering dogwood in Catoctin Mountain Park. Plant Disease 80:310–312.

Smith, B. 1992. A tree grows in Brooklyn. New York: Random House.

Srivastava, V., V.C. Griess, and M.A. Keena. 2020. Assessing the potential distribution of Asian gypsy moth in Canada: A comparison of two methodological approaches. Scientific Reports 10:1–10.

Tisserat, N., W. Cranshaw, M.L. Putnam, J. Pscheidt, C.A. Leslie, M. Murray, J. Hoffman, Y. Barkley, K. Alexander, and S.J. Seybold. 2011. Thousand cankers disease is widespread in black walnut in the western United States. Plant Health Progress 12:35.

Trigiano, R.N., G. Caetano-Anollés, B.J. Bassam, T. Mark, and R.N. Trigiano. 1995. DNA amplification fingerprinting provides evidence that *Discula destructiva*, the cause of dogwood anthracnose in North America, is an introduced pathogen. Mycologia 87:490–500.

Wiggins, G.J., J.F. Grant, P.L. Lambdin, P. Merten, K.A. Nix, D. Hadziabdic, and M.T. Windham. 2014. Discovery of walnut twig beetle, *Pityophthorus juglandis*, associated with forested black walnut, *Juglans nigra*, in the eastern U.S. Forests 5:1185–1193.

Williams, C.E., and W.J. Moriarity. 1999. Occurrence of flowering dogwood (*Cornus florida* L.), and mortality by dogwood anthracnose (*Discula destructiva* Redlin), on the Northern Allegheny Plateau. Journal of the Torrey Botanical Society 126:313–319.

Zahiri, R., B.C. Schmidt, A. Schintlmeister, R.V. Yakovlev, and M. Rindoš. 2019. Global phylogeography reveals the origin and the evolutionary history of the gypsy Moth (Lepidoptera, Erebidae). Molecular Phylogenetics and Evolution 137:1–13.

Zerillo, M.M., J. Ibarra Caballero, K. Woeste, A.D. Graves, C. Hartel, J.W. Pscheidt, J. Tonos, et al. 2014. Population structure of *Geosmithia morbida*, the causal agent of thousand cankers disease of walnut trees in the United States. PLoS ONE 9(11): e112847.

## The Next in Line

Albright, T.P., D.P. Anderson, N.S. Keuler, S.M. Pearson, and M.G. Turner. 2009. The spatial legacy of introduction: *Celastrus orbiculatus* in the southern Appalachians, USA. Journal of Applied Ecology 46:1229–1238.

Aronson, M.F.J., and S.N. Handel. 2011. Deer and invasive plant species suppress forest herbaceous communities and canopy tree regeneration. Natural Areas Journal 31:400–407.

Averill, K.M., D.A. Mortensen, E.A.H. Smithwick, and E. Post. 2016. Deer feeding selectivity for invasive plants. Biological Invasions 18:1247–1263.

Baiser, B., J. Lockwood, D. La Puma, and M. Aronson. 2008. A perfect storm: Two ecosystem engineers interact to degrade deciduous forests of New Jersey. Biological Invasions 10:785–795.

Barringer, L., and C.M. Ciafré. 2020. Worldwide feeding host plants of spotted lanternfly, with significant additions from North America. Environmental Entomology 49:999–1011.

Bialic-Murphy, L., N.L. Brouwer, and S. Kalisz. 2020. Direct effects of a non-native invader erode native plant fitness in the forest understory. Journal of Ecology 108:189–198.

Cavers, P.B., M.I. Heagy, and R.F. Kokron. 1979. The biology of Canadian weeds. 35. *Alliaria petiolata* (M. Bieb) Cavara and Grande. Canadian Journal of Plant Science 59:217–229.

Cheplick, G.P. 2006. A modular approach to biomass allocation in an invasive annual (*Microstegium vimineum*: Poaceae). American Journal of Botany 93:539–545.

Cincotta, C.L., J.M. Adams, and C. Holzapfel. 2009. Testing the enemy release hypothesis: A comparison of foliar insect herbivory of the exotic Norway maple (*Acer platanoides* L.) and the native sugar maple (*A. saccharum* L.). Biological Invasions 11:379–388.

Cipollini, D., C.M. Rigsby, and E.K. Barto. 2012. Microbes as targets and mediators of allelopathy in plants. Journal of Chemical Ecology 38:714–727.

Cole, P.G., and J.F. Weltzin. 2005. Light limitation creates patchy distribution of an invasive grass in eastern deciduous forests. Biological Invasions 7:477–488.

Collier, M.H., and J.L. Vankat. 2002. Diminished plant richness and abundance below *Lonicera maackii*, an invasive shrub. American Midland Naturalist 147:60–71.

Côté, S.D., T.P. Rooney, J.-P. Tremblay, C. Dussault, and D.M. Waller. 2004. Ecological impacts of deer overabundance. Annual Review of Ecology, Evolution, and Systematics 35:113–147.

Deering, R.H., and J.L. Vankat. 1999. Forest colonization and developmental growth of the invasive shrub *Lonicera maackii*. American Midland Naturalist 141:43–50.

Dillenburgh, L.R., D.F. Whigham, A.H. Teramura, and I.N. Forseth. 1993. Effects of vine competition on availability of light, water, and nitrogen to a tree host (*Liquidambar styraciflua*). American Journal of Botany 80:244–252.

Dodds, K.J., and D.A. Orwig. 2011. An invasive urban forest pest invades natural environments—Asian long horned beetle in northeastern US hardwood forests. Canadian Journal of Forest Research 41:1729–1742.

Dreyer, G.D., L.M. Baird, and C. Fickler. 1987. *Celastrus scandens* and *Celastrus orbiculatus*: Comparisons of reproductive potential between a native and an introduced woody vine. Bulletin of the Torrey Botanical Club 114:260–264.

Driscoll, A.G., N.F. Angeli, D.L. Gorchov, Z. Jiang, J. Zhang, and C. Freeman. 2016. The effect of treefall gaps on the spatial distribution of three invasive plants in a mature upland forest in Maryland. Journal of the Torrey Botanical Society 143:349–358.

Droste, T., S.L. Flory, and K. Clay. 2010. Variation for phenotypic plasticity among populations of an invasive exotic grass. Plant Ecology 207:297–306.

Ellsworth, J.W., R.A. Harrington, and J.H. Fownes. 2016. Growth and gas exchange of *Celastrus orbiculatus* seedlings in sun and shade. American Midland Naturalist 151:233–240.

Fike, J., and W.A. Niering. 1999. Four decades of old field vegetation development and the role of *Celastrus orbiculatus* in the northeastern United States. Journal of Vegetation Science 10:483–492.

Flory, S.L., and K. Clay. 2010. Non-native grass invasion suppresses forest succession. Oecologia 164:1029–1038.

Gibson, D.J., G. Spyreas, and J. Benedict. 2002. Life history of *Microstegium vimineum* (Poaceae), an invasive grass in southern Illinois. Journal of the Torrey Botanical Society 129:207–219.

Gómez-Aparicio, L., and C.D. Canham. 2008. Neighbourhood analyses of the allelopathic effects of the invasive tree *Ailanthus altissima* in temperate forests. Journal of Ecology 96:447–458.

Gorchov, D.L., B. Blossey, K.M. Averill, A. Dávalos, J.M. Heberling, M.A. Jenkins, S. Kalisz, et al. 2021. Differential and interacting impacts of invasive plants and white-tailed deer in eastern U. S. Forests. Biological Invasions. 23:2711–27.

Gorchov, D.L., E. Thompson, J. O'Neill, D. Whigham, and D.A. Noe. 2011. Treefall gaps required for establishment, but not survival, of invasive *Rubus phoenicolasius* in deciduous forest, Maryland, USA. Plant Species Biology 26:221–234.

Gorchov, D.L., and D.E. Trisel. 2003. Competitive effects of the invasive shrub, *Lonicera maackii* (Rupr.) Herder (Caprifoliaceae), on the growth and survival of native tree seedlings. Plant Ecology 166:13–24.

Hale, A.N., L. Lapointe, and S.C. Kalisz. 2016. Invader disruption of belowground plant mutualisms reduces carbon acquisition and alters allocation patterns in a native forest herb. New Phytologist 209:542–549.

Hale, A.N., S.J. Tonsor, and S. Kalisz. 2011. Testing the mutualism disruption hypothesis: Physiological mechanisms for invasion of intact perennial plant communities. Ecosphere 2:1–15.

Hartman, K.M., and B.C. McCarthy. 2004. Restoration of a forest understory after the removal of an invasive shrub, Amur honeysuckle (*Lonicera maackii*). Restoration Ecology 12:154–165.

Horton, J.L., and H.S. Neufeld. 1998. Photosynthetic responses of *Microstegium vimineum* (Trin.) A. Camus, a shade-tolerant, $C_4$ grass, to variable light environments. Oecologia 114:11–19.

Hoven, B.M., D.L. Gorchov, K.S. Knight, and V.E. Peters. 2017. The effect of emerald ash borer-caused tree mortality on the invasive shrub Amur honeysuckle and their combined effects on tree and shrub seedlings. Biological Invasions 19:2813–2836.

Hutchinson, T.F., and J.L. Vankat. 1998. Landscape structure and spread of the exotic shrub *Lonicera maackii* (Amur honeysuckle) in southwestern Ohio Forests. American Midland Naturalist 139:383–390.

Kuhman, T.R., S.M. Pearson, and M.G. Turner. 2011. Agricultural land-use history increases non-native plant invasion in a southern Appalachian forest a century after abandonment. Canadian Journal of Forest Research 41:920–929.

Kuhman, T.R., S.M. Pearson, and M.G. Turner. 2013. Why does land-use history facilitate non-native plant invasion? A field experiment with *Celastrus orbiculatus* in the southern Appalachians. Biological Invasions 15:613–626.

Ladwig, L.M., and S.J. Meiners. 2009. Impacts of temperate lianas on tree growth in young deciduous forests. Forest Ecology and Management 259:195–200.

Ladwig, L.M., and S.J. Meiners. 2010. Liana host preference and implications for deciduous forest regeneration. Journal of the Torrey Botanical Society 137:103–112.

Ladwig, L.M., and S.J. Meiners 2010. Spatio-temporal dynamics of lianas during 50 years of succession to temperate forest. Ecology 91:671–680.

Lankau, R.A. 2011. Intraspecific variation in allelochemistry determines an invasive species' impact on soil microbial communities. Oecologia 165:453–463.

Lawrence, J.G., A. Colwell, and O.J. Sexton. 1991. The ecological impact of allelopathy in *Ailanthus altissima* (Simaroubaceae). American Journal of Botany 78:948–958.

Leicht, S.A., and J.A. Silander. 2006. Differential responses of invasive *Celastrus orbiculatus* (Celastraceae) and native *C. scandens* to changes in light quality. American Journal of Botany 93:972–977.

Loomis, J.D., S.F. Matter, and N. Guy. 2015. Effects of invasive Amur honeysuckle (*Lonicera maackii*) and white-tailed deer (*Odocoileus virginianus*) on survival of sugar maple seedlings in a southwestern Ohio forest. American Midland Naturalist 174:65–73.

Martin, P.H. 1999. Norway maple (*Acer platanoides*) invasion of a natural forest stand: Understory consequence and regeneration pattern. Biological Invasions 1:215–222.

Meiners, S.J. 2007. Apparent competition : An impact of exotic shrub invasion on tree regeneration. Biological Invasions 9:849–855.

Miller, K.E., and D.L. Gorchov. 2004. The invasive shrub, *Lonicera maackii*, reduces growth and fecundity of perennial forest herbs. Oecologia 139:359–375.

Miyake, T., and T. Yahara. 1998. Why does the flower of *Lonicera japonica* open at dusk? Canadian Journal of Botany 76:1806–1811.

Roberts, K.J., and R.C. Anderson. 2005. Effect of garlic mustard [*Alliaria petiolata* (Beib, Cavarra & Grande)] extracts on plants and arbuscular mycorrhizal (AM) fungi. American Midland Naturalist 146:146–152.

Robertson, D.J., M.C. Robertson, and T. Tague. 1994. Colonization dynamics of four exotic plants in a northern piedmont natural area. Bulletin of the Torrey Botanical Club 121:107–118.

Rooney, T.P. 2001. Deer impacts on forest ecosystems: A North American perspective. Forestry 74:201–208.

Schmidt, K.A., and C.J. Whelan. 1999. Effects of exotic *Lonicera* and *Rhamnus* on songbird nest predation. Conservation Biology 13:1502–1506.

Shen, X., N.A. Bourg, W.J. McShea, and B.L. Turner. 2016. Long-term effects of white-tailed deer exclusion on the invasion of exotic plants: A case study in a mid-Atlantic temperate forest. PLoS ONE 11(3): e0151825.

Steward, A.M., S.E. Clemants, and G. Moore. 2003. The concurrent decline of the native *Celastrus scandens* and spread of the non-native *Celastrus orbiculatus* in the New York City metropolitan area. Journal of the Torrey Botanical Society 130:143–146.

Tibbetts, T.J., and F.W. Ewers. 2000. Root pressure and specific conductivity in temperate lianas: Exotic *Celastrus orbiculatus* (Celastraceae) vs. native *Vitis riparia* (Vitaceae). American Journal of Botany 87:1272–1278.

U.S. Department of Agriculture. 2013. Pest alert: Spotted lanternfly. Pest Alert DACS-P-01863.

Williams, V.R.J., and H.F. Sahli. 2016. A comparison of herbivore damage on three invasive plants and their native congeners: Implications for the enemy release hypothesis. Castanea 81:128–137.

Wyckoff, P.H., and S.L. Webb. 1996. Understory influence of the invasive Norway maple (*Acer platanoides*). Bulletin of the Torrey Botanical Club 123:197–205.

## Accumulating Impacts—Putting It All Together

Ambler, M.A. 1965. Seven alien plant species. William L. Hutcheson Memorial Forest Bulletin 2:1–8.

Baiser, B., J. Lockwood, D. La Puma, and M. Aronson. 2008. A perfect storm: Two ecosystem engineers interact to degrade deciduous forests of New Jersey. Biological Invasions 10:785–795.

Bard, G.E. 1952. Secondary succession on the Piedmont of New Jersey. Ecological Monographs 22:195–215.

Buell, M.F., H.F. Buell, and J.A. Small. 1954. Fire in the history of Mettler's Woods. Bulletin of the Torrey Botanical Club 81:253–255.

Davison, S.E., and R.T.T. Forman. 1982. Herb and shrub dynamics in a mature oak forest: A thirty-year study. Bulletin of the Torrey Botanical Club 109:64–73.

Frei, K.R., and D.E. Fairbrothers. 1963. Floristic study of the William L. Hutcheson Memorial Forest (New Jersey). Bulletin of the Torrey Botanical Club 90:338–355.

Monk, C.D. 1957. Plant communities of Hutcheson Memorial Forest based on shrub distribution. Bulletin of the Torrey Botanical Club 84:198–206.

Monk, C.D. 1961. Past and present influences on reproduction in the William L. Hutcheson Memorial Forest, New Jersey. Bulletin of the Torrey Botanical Club 88:167–175.

Wales, B.A. 1972. Vegetation analysis of north and south edges in a mature oak–hickory forest. Ecological Monographs 42:451–471.

## Conclusion: Protecting Our Forests' Future

Allen, E., M. Noseworthy, and M. Ormsby. 2017. Phytosanitary measures to reduce the movement of forest pests with the international trade of wood products. Biological Invasions 19:3365–3376.

Beaury, E.M., E.J. Fusco, J.M. Allen, and B.A. Bradley. 2021. Plant regulatory lists in the United States are reactive and inconsistent. Journal of Applied Ecology 58:1957–1966.

Beaury, E.M., M. Patrick, and B.A. Bradley. Invaders for sale: The ongoing spread of invasive species by the plant trade industry. Frontiers in Ecology and the Environment 19:550–556.

Cleavitt, N.L., J.J. Battles, T.J. Fahey, and N.S. Van Doorn. 2021. Disruption of the competitive balance between foundational tree species by interacting stressors in a temperate deciduous forest. Journal of Ecology 109:2754–2768.

Coop, J.D., S.A. Parks, C.S. Stevens-Rumann, S.D. Crausbay, P.E. Higuera, M.D. Hurteau, A. Tepley, et al. 2020. Wildfire-driven forest conversion in western North American landscapes. BioScience 70:659–673.

Dey, D.C., B.O. Knapp, M.A. Battaglia, R.L. Deal, J.L. Hart, K.L. O'Hara, C.J. Schweitzer, and T.M. Schuler. 2019. Barriers to natural regeneration in temperate forests across the USA. New Forests 50:11–40.

Diagne, C., B. Leroy, A.-C. Vaissière, R.E. Gozlan, D. Roiz, I. Jarić, J.-M. Salles, C.J.A. Bradshaw, and F. Courchamp. 2021. High and rising economic costs of biological invasions worldwide. Nature 592:571–576.

Fridley, J.D. 2008. Of Asian forests and European fields: Eastern U.S. plant invasions in a global floristic context. PLoS ONE 3(11): e3630.

Hagerman, S., and R. Kozak. 2021. Disentangling the social complexities of assisted migration through deliberative methods. Journal of Ecology 109:2309–2316.

Hennon, P.E., S.J. Frankel, A.J. Woods, J.J. Worrall, T.D. Ramsfield, P.J. Zambino, D.C. Shaw, et al. 2021. Applications of a conceptual framework to assess climate controls of forest tree diseases. Forest Pathology 51(6): e12719.

Liebhold, A.M., E.G. Brockerhoff, and M.A. Nuñez. 2017. Biological invasions in forest ecosystems: A global problem requiring international and multidisciplinary integration. Biological Invasions 19:3073–3077.

Mccullough, D.G., T.T. Work, J.F. Cavey, A.M. Liebhold, and D. Marshall. 2006. Interceptions of nonindigenous plant pests at US ports of entry and border crossings over a 17-year period. Biological Invasions 8:611–630.

Pimentel D., M. Pimentel, and A. Wilson. 2008. Plant, animal, and microbe invasive species in the United States and world. In: Biological Invasions, ed. W. Nentwig, 315–330. Ecological Studies (Analysis and Synthesis), vol. 193. Berlin: Springer.

Thompson, J.R., D.N. Carpenter, C.V. Cogbill, and D.R. Foster. 2013. Four centuries of change in Northeastern United States forests. PLoS ONE 8(9): e72540.

Wason, J.W., M. Dovciak, C.M. Beier, and J.J. Battles. 2017. Tree growth is more sensitive than species distributions to recent changes in climate and acidic deposition in the northeastern United States. Journal of Applied Ecology 54:1648–1657.

# Index

Page numbers followed by f and m refer to figures and maps, respectively.

Printed in the USA
CPSIA information can be obtained
at www.ICGtesting.com
CBHW020307260324
5886CB00001B/67

9 781501 771262